T0135763

Diss. ETH No. 20657

Integral Methods for Quadratic Programming: Theory and Implementation

A dissertation submitted to
ETH ZÜRICH

for the degree of
DOCTOR OF SCIENCES

presented by
YVES DOMINIQUE BRISE
Dipl. Informatik-Ing. ETH
born July 27, 1980
citizen of Allschwil BL, Switzerland

accepted on the recommendation of
Prof. Dr. Emo Welzl, examiner
Dr. Bernd Gärtner, co-examiner
Prof. Dr. Friedrich Eisenbrand, co-examiner

2012

Bibliografische Information der Deutschen Nationalbibliothek

Die Deutsche Nationalbibliothek verzeichnet diese Publikation in der
Deutschen Nationalbibliografie; detaillierte bibliographische Daten sind
im Internet über http://dnb.d-nb.de abrufbar.

Diese Publikation ist bei der Schweizer Nationalbibliothek und bei der
Bibliothek der ETH Zürich hinterlegt. Weitere Information sind im
Internet unter http://e-collection.library.ethz.ch abrufbar.

ISBN 978-3-8325-3366-3

Logos Verlag Berlin GmbH
Comeniushof, Gubener Str. 47,
10243 Berlin
Tel.: +49 (0)30 42 85 10 90
Fax: +49 (0)30 42 85 10 92
INTERNET: http://www.logos-verlag.de

Dedicated to my first child,
who is waiting to be born;
and no less to Werner,
who would have been one of
the few people to actually read this book.

Abstract

The contributions of this thesis are twofold. We show two theoretical results that are both related to *quadratic programming*.

The first one concerns the abstract optimization framework of *violator spaces* and the randomized procedure called *Clarkson's algorithm*, which is associated with solving violator spaces. Historically, Clarkson's algorithm was developed to solve linear programs, and provided the earliest practical linear-time algorithm for linear programs. The underlying concept of the problems solvable by this algorithm was later expanded to *LP-type problems*, and finally to violator spaces. Quadratic programming is also an LP-type problem. In a nutshell, the algorithm randomly samples from a set of constraints, computes an optimal solution subject to these constraints, and then checks whether the ignored constraints agree with the solution. If not, some form of re-sampling occurs, until an optimal solution is found that satisfies all constraints. Originally, to make the analysis go through, there used to be a preliminary test whether the random sample is good in some sense. We show that this test is not necessary and we give evidence that the modified version of Clarkson's approach is the easiest version that can still be analyzed successfully.

The second contribution concerns quadratic programming more directly. It is well-known that a simplex like procedure can be applied to quadratic programming – similar to the simplex algorithm for linear programming. The main computational effort in this algorithm comes from solving a series of linear equation systems that change gradually. We develop a method that allows for efficiently solving these systems under the assumption that (i) we want to do exact computations using some arbitrary precision number type, and (ii) the input may be sparse, and that should be exploited. In particular, the tool of choice is the LU factorization of the matrix to invert, and we call our algorithm the *integral LU factorization*. We also give an algorithm to update the factorization subject to low rank changes of the original matrix, which covers the gradual changes during simplex iterations that we mentioned.

Last but not least, a considerable portion of the work included in this thesis was devoted to implementing the integral LU factorization in the framework of the existing quadratic programming solver in the Computational Geometry Algorithms Library (CGAL). In the last two chapters we describe our implementation and the experimental results we obtained.

Zusammenfassung

Der Beitrag dieser Arbeit ist zweigeteilt. Wir zeigen zwei theoretische Resultate, die beide im Zusammenhang mit *quadratischen Programmen* stehen. Das erste betrifft die abstrakte Klasse von Optimierungsproblemen, die wir *Verletzerräume* nennen; insbesondere *Clarkson's Algorithmus*, der dazu verwendet wird, um Verletzerräume zu lösen. Historisch gesehen wurde Clarkson's Algorithmus entwickelt, um lineare Programme zu lösen, und stellt den ersten praktikablen Linearzeit-Algorithmus für ebendiese dar. Die zugrunde liegende Beschreibung der Probleme, die durch diesen Algorithmus lösbar sind, wurde später auf *LP-Typ Probleme*, und schliesslich auf Verletzerräume ausgedehnt. Quadratische Programme gehören auch zu den LP-Typ Problemen. Kurz gesagt, macht der Algorithmus Folgendes: Aus der Menge der Nebenbedingungen wird eine zufällige Auswahl getroffen. Dann wird eine optimale Lösung unter diesen Nebenbedingungen berechnet und überprüft, ob eine der bislang vernachlässigten Nebenbedingungen diese Lösung verletzt. Falls dies der Fall ist, wird die zufällige Auswahl modifiziert, und eine neue Lösung berechnet. Dies wird so lange iteriert, bis alle Nebenbedingungen durch die aktuelle Lösung erfüllt sind. Ursprünglich gab es bei jeder zufälligen Auswahl einen Test, der überprüft hat, ob die Auswahl in einem bestimmten Sinne gut sei.

Dies wurde getan, um die Laufzeit erfolgreich zu analysieren. Wir zeigen, dass diese Tests nicht nötig sind, und führen aus, dass der resultierende Algorithmus vermutlich die einfachste Form von Clarkson's Algorithmus darstellt, die noch erfolgreich analysiert werden kann.

Der zweite Beitrag betrifft quadratische Programme direkter. Es ist bekannt, dass ein Simplex-artiger Algorithmus auf quadratische Programme angewendet werden kann, analog zum Simplex Algorithmus für lineare Programme. Der grösste Berechnungsaufwand bei dieser Methode entsteht aus der Notwendigkeit, eine Reihe von linearen Gleichungssystemen zu lösen, die sich graduell verändern. Wir entwickeln eine Methode, die diese Gleichungssysteme effizient löst unter der Annahme, dass wir (*i*) exakte Berechnungen mit einem Zahlentyp durchführen wollen, der eine beliebige Präzision erlaubt, und (*ii*) die Eingabe dünn sein kann, was ausgenützt werden soll. Konkret stützen wir uns dabei auf die LU Zerlegung der Matrix, die es zu invertieren gilt, und wir nennen unseren Algorithmus die *integrale LU Zerlegung*. Zusätzlich geben wir einen Algorithmus an, mit dem man die Zerlegung einer Matrix aktualisieren kann, angenommen die Matrix habe sich durch einen additiven Term niedrigen Ranges verändert. Dies verwenden wir um die graduellen Veränderungen während eines Iterationsschrittes des Simplex Algorithmus zu behandeln.

Zu guter Letzt besteht ein beträchtlicher Teil des Aufwandes, dessen Resultate in dieser Arbeit besprochen werden, daraus, die integrale LU Zerlegung im Rahmengerüst des Lösers für quadratische Programme in der Computational Geometry Algorithms Library (CGAL) zu implementieren. In den letzten beiden Kapiteln besprechen wir diese Implementierung und die experimentellen Resultate, die wir erhalten haben.

Acknowledgments

First and foremost my thanks and acknowledgements go to my supervisor Bernd Gärtner and my professor Emo Welzl for their continuing support, their patience, and valuable input during all stages of the thesis. Many thanks go to Friedrich Eisenbrand as well, for inviting me to Lausanne and for agreeing to review my thesis.

Second and very importantly I would like to thank the colleagues who have kindly agreed to help me proofread parts of the thesis. In alphabetical order they are Timon Hertli, Martin Jaggi, Robin Moser, Andreas Razen, and Sebastian Stich.

During the time of my thesis it has been a great pleasure to work as part of the CGAL development team. I enjoyed meeting those engaging and welcoming people, in particular: Eric Berberich, Mikhail Bogdanov, Yacine Bouzidi, Manuel Caroli, Guillaume Damiand, Olivier Devillers, Pavel Emeliyanenko, Efi Fogel, Bernd Gärtner, Marc Glisse, Michael Hemmer, Michael Hoffmann, Menelaos Karavelas, Alexander Kobel, Sébastien Loriot, Luis Peñaranda, Sylvain Pion, Marc Pouget, Laurent Rineau, Fabrice Rouillier, Stéphane Tayeb, Monique Teillaud, and Mariette Yvinec.

Former members who have worked on the CGAL project – and to who I feel a connection even though I have never met them – include Kaspar Fischer, Sven Schönherr, and Frans Wessendorp. They have worked hard on the quadratic programming solver, so I could take up their work and try to make it even better.

I am indebted to my colleagues from the GREMO group at ETH, for scientific discussions as well as for the occasional game session or the Saturday night out: Robert Berke, Tobias Christ, Andrea Francke, Bernd Gärtner, Heidi Gebauer, Anna Gundert, Franziska Hefti, Timon Hertli, Michael Hoffmann, Martin Jaggi, Vincent Kusters, Robin Moser, Gabriel Nivasch, Andreas Razen, Leo Rüst, Andrea Salow, Piotr Sankowski, Dominik Scheder, Eva Schuberth, Sebastian Stich, Marek Sulovský, Tibor Szabó, Patrick Traxler, Hemant Tyagi, Uli Wagner, and Philipp Zumstein.

Two former members of the GREMO group – who I had the pleasure of doing a semester thesis with as a student – are Joachim Giesen and Dieter Mitsche. Similarly, it was my great pleasure to supervise Manuel Wettstein's thesis on the Clarkson dimension together with Bernd Gärtner.

Finally, my mother Sirkka, my wife Lily, my godchildren Arthur, Iida, and Linus, the other members of my family and my friends helped me find the peace of mind and the serenity that are indispensable for having the resolve to finish such a big project – and that wasn't always the easiest job in the world. Many thanks to them!

Contents

*The left hand now knows what the
right hand is doing.*

George W. Bush

Introduction

1.1. An Optimization Problem

1.1.1. Quadratic Programming

Optimization problems from different fields can be formulated as
quadratic programs, or as their even better studied specialization called
linear programs. A quadratic program asks to minimize a quadratic
objective function of several variables subject to a set of linear constraints on these variables. In its most general form it can be written

as follows,

$$(\text{QP}) \qquad \min \qquad c^T x + x^T D x$$

$$\text{s.t.} \qquad Ax \gtreqless b \tag{1.1}$$

$$\ell \leq x \leq u,$$

where A is an $m \times n$-matrix, D is an $n \times n$-matrix, b is an m-vector, c is an n-vector and ℓ, u are n-vectors of *bounds* (where the entries $+\infty$ and $-\infty$ may occur). The matrix A is called the *constraint matrix*. The symbol $\gtreqless$ indicates that each of the three relations $\leq$, $=$, or $\geq$ is admissible for a particular constraint. The n-vector x is the *solution vector*, and consists of the variables that we have to find optimal values for.

A solution vector $x^* = (x_1^*, \ldots, x_n^*)^T$ is called *feasible solution* if it satisfies all the constraints and bounds. If no feasible solution exists, the problem is called *infeasible*. The region that is defined by the constraint (in)equalities is called the *feasible region*. If the objective function $f(x) := c^T x + x^T D x$ is bounded from below, we say that the problem is *bounded*, otherwise we say that it is *unbounded*.

We will also consider the *equality constrained* formulation, which is sometimes called the *standard* formulation

$$(\text{EQP}) \qquad \min \qquad c^T x + x^T D x$$

$$\text{s.t.} \qquad Ax = b \tag{1.2}$$

$$x \geq 0,$$

and the *unconstrained* formulation

$$(\text{UQP}) \qquad \min \qquad c^T x + x^T D x$$

$$\text{s.t.} \qquad Ax = b. \tag{1.3}$$

In a certain sense, all three forms are equivalent and can be converted into each other (possibly involving a change in the number of variables and constraints). From an algorithmic point of view, the unconstrained version UQP is strictly easier to solve using the method of *Lagrange multipliers*. The different forms will serve us to highlight different aspects of the problem description and the solution process. Also note that the formulations for maximization are easily obtained by multiplying $f(x)$ by -1. Throughout the thesis we will always state optimization tasks as minimization problems.

If $D = 0$ then we have a *linear program* at hand. In the other case, $D \neq 0$, we only consider *positive semidefinite* matrices in this thesis; that is, $x^T D x \geq 0$ holds for all vectors x. This condition is equivalent to saying that the objective function is convex (or strictly convex if D is positive definite). In that case any local optimum of $f(x)$ is also a global optimum. While there exist (weakly) polynomial algorithms for semidefinite quadratic programming [85, 155], it is *NP*-hard to find the global minimum of a non-convex QP [127, 144]. Furthermore, finding the local optimum of a non-convex QP – and under certain conditions even checking local optimality – is *NP*-hard [109, 118]. Sometimes, *quadratically constrained quadratic programs* are considered, where the constraints on the variables may themselves be quadratic. This variation is also *NP*-hard, because the constraint $x_1(x_1 - 1) = 0$ requires the solution to attain a discrete value $x_1 \in \{0, 1\}$. This, in turn, means that quadratically constrained QPs are a generalization of 0-1 integer programs, which belong to Karp's 21 *NP*-complete problems [78].

1.1.2. Applications

The array of applications for quadratic programming is vast, and we will only give a brief overview. Most important, let us point out the

website *A Quadratic Programming Page*[1], that is actively maintained
by Gould and Toint. It contains an up-to-date BibTeX collection
of almost one thousand papers revolving around quadratic programming, including many applications. To name a few areas, let us mention *portfolio analysis* [20, 112, 32, 131, 97, 96], *VLSI design* [16, 87,
160, 79, 51, 52, 83], *discrete-time stabilization* [17, 138, 93, 124], *optimal and fuzzy control* [77, 89, 15, 95, 104, 84, 81, 72], *finite impulse
control* [114, 90, 91, 103], *optimal power flow* [105, 19, 146, 106, 111,
116], *economic dispatch* [50, 68, 73, 9, 24, 28, 121, 107], and *geometric
optimization*. The latter problem class has been treated and surveyed
extensively by Schönherr in his PhD thesis [128]. In this exposition –
especially in the implementation part – we continue his work.

   Some of these applications mentioned arise from the extension of
quadratic programming to non-linear (and generally non-quadratic)
optimization. The following optimization problem NLP is only restricted by the condition that the objective function $g : \mathbb{R}^n \to \mathbb{R}$
and the constraint functions b and c have to be twice continuously
differentiable. The nonlinear problem

$$
\begin{aligned}
\text{(NLP)} \quad \min \quad & g(x) \\
\text{s.t.} \quad & b(x) \geq 0 \\
& c(x) = 0
\end{aligned}
\tag{1.4}
$$

can be solved by iterating through a series of approximate solution
vectors x_k. At every step, the search direction d_k by which we
change the iterate is determined by a quadratic programming sub-
problem. This method is known as *sequential quadratic programming* (SQP) [134, 64].

---

[1] http://www.numerical.rl.ac.uk/qp/qp.html, see also [66].

Let us illustrate two examples of quadratic programming problems in the following paragraphs, which will outline one of the main motivations for this thesis.

Problem 1.5 (Smallest Enclosing Ball).
Given a sets of points $P = \{p_1, \ldots, p_n\} \in \mathbb{R}^d$, determine the smallest ball that contains all the points.

If we define the $d \times n$-matrix $C := (p_1, \ldots, p_n)$ to be the matrix that contains the coordinates of the points as its columns, we can write Problem 1.5 as the following quadratic program,

$$(\text{SEB}) \qquad \min \qquad x^T C^T C x - \sum_{i=1}^{n} p_i^T p_i x_i$$

$$\text{s.t.} \qquad \sum_{i=1}^{n} x_i = 1 \qquad\qquad (1.6)$$

$$x \geq 0.$$

It is not trivial to see but proved in Theorem 3.1 of [128] that any optimal solution $x^* = (x_1^*, \ldots, x_n^*)^T$ to this problem determines the center c of the smallest enclosing ball,

$$c = \sum_{i=1}^{n} x_i^* p_i.$$

Furthermore, the squared radius of the ball is given by the negative value of the objective function at x^*.

We notice that the quadratic part of the objective function, $C^T C$, is fully dense[2]. Even if it were not, there is always a translation of P,

---

[2] *Dense* means that most of the entries of the matrix are nonzero. By contrast, *sparse* means the opposite, namely that most of the entries are zero. Note that there is no formal definition of what "most" means in this context.

such that all its coordinates are nonzero (and positive). Obviously, such a translate P' results in a translated, but otherwise identical, solution. So, except for specifically constructed cases, we have to assume dense input. However, the matrix $D := C^T C$ also has the convenient property that its rank is at most d. For fixed values of d, this makes the problem tractable by Schönherr's simplex algorithm, even if the number of points is large – typically $n \gg d$. This is due to Theorem 2.6 of [128] (restated as Theorem 3.5 in this thesis).

We are going to describe that algorithm in more detail in Chapter 3. Sven Schönherr developed it together with Bernd Gärtner [58], and it was implemented in the *Computational Geometry Algorithms Library (CGAL)*[3] by the aforementioned, Kaspar Fischer, and Franz Wessendorp. A major part of the present thesis deals with extending that implementation. We will get to that in Chapters 5 and 6. For now, let us continue with our introduction.

The important point about the smallest enclosing ball example is, that Schönherr's simplex algorithm is specifically tuned to this type of application that often arises in computational geometry optimization problems. More precisely, it profits from $\min\{m, n\}$ being small, but it is insensitive to the occurrence of nonzero entries in the input. For completeness, we note that other efficient algorithms exist for the smallest enclosing ball problem: a randomized algorithm by Welzl [147], approximation algorithms [157, 86], and a combinatorial exact algorithm [53].

Now let us come to the second problem we announced earlier on. Consider a chemical plant. The plant can produce n different products that sell at a price p_j each and come at a cost of c_j. The contribution

---

[3] http://www.cgal.org/

margin for each product is defined as $d_j := p_j - c_j$. The plant has
a set of m machines whose workings are described by the production
coefficients $a_{i,j}$. For each product j the value $a_{i,j}$ describes how much
of the capacity of machine i is used to produce a unit of product j.
The production coefficients can be aggregated in a matrix

$$A = \begin{pmatrix} a_{1,1} & \cdots & a_{1,n} \\ \vdots & & \vdots \\ a_{m,1} & \cdots & a_{m,n} \end{pmatrix}.$$

Each machine also comes with a maximal capacity b_i. The managers
of the plant would like to maximize their profits through maximizing
the sum of production margins.

Problem 1.7 (Maximizing Contribution Margins).
*Given the vector of contribution margins $d = (d_1, \ldots, d_n)$, the vector
of maximal capacities $b = (b_1, \ldots, b_m)$, and the coefficient matrix A,
minimize $-d^T x$.*

Maximizing $d^T x$ is of course equivalent to minimizing $-d^T x$. We
readily arrive at the following linear programming formulation,

$$
\begin{aligned}
\text{(MCM)} \qquad \min \quad & -d^T x \\
\text{s.t.} \quad & Ax \leq b \\
& x \geq 0.
\end{aligned}
\qquad (1.8)
$$

The crucial point here is that for a large plant that produces a large
number of products, the matrix A can be expected to be extremely
sparse, because a particular machine (or process) is likely to be able to
produce only a small number of products. This example illustrated a
setting that is typically encountered in large scale optimization prob-

lems from operations research.

Of course, this is not a proper quadratic program, but actual large-scale applications for quadratic programming often come from SQP[4] formulations, which do not lend themselves to an easy description. For another relatively simple example with a (proper) sparse quadratic programming formulation consider the problem of *tabular data protection*; also known as *statistical exposure control* [25].

Above examples outline one of the main goals of this thesis. Schönherr's implementation is not suited for the operations research setting, where both m and n are large, usually in the hundreds if not thousands. Implementations that are able to solve such large problems need to be tackled differently. We need to take advantage of the large number of zeros in the problem input. To see how it is possible to incorporate this into Schönherr's approach, let us first give a general overview of the methods that have been developed to solve QPs.

1.1.3. Solution Methods

Essentially there are two different classes of algorithms to solve a quadratic program in its general form (1.1): *interior-point* and *active-set* methods. Another method that can be employed is the *trust-region* approach; for a survey see [156]. We will restrict our discussion to the former two. Also, as we have said earlier, we only consider convex problems, i.e., D is assumed to be positive definite.

At their heart both of these methods – active-set as well as interior-point – are based on the Karush-Kuhn-Tucker conditions for convex optimization (see any standard textbook on convex optimization, e.g., [21]). Without going into too much detail for now, both methods

---

[4] Recall that SQP stands for *sequential quadratic programming*.

go through a series of iterations in each of which a linear system of equations has to be solved. We will call this the KKT system. The relevant matrix looks as follows,

$$
\begin{pmatrix} 0 & A_k \\ A_k^T & D_k \end{pmatrix}, \tag{1.9}
$$

where A_k and D_k are sub-matrices of A and D respectively.

Interior-point methods maintain an approximate solution that lies strictly within the feasible region. In each iteration we have to solve a KKT system that depends on *all* constraints and variables, i.e., $D_k = D + \Delta^{(k)}$ and $A_k = A$, where the diagonal matrix $\Delta^{(k)}$ changes from iteration to iteration. The number of iterations, however, is usually low and almost independent of the problem size.

By contrast, active-set methods try to reduce the amount of work necessary during each iteration by reducing the size of the KKT system considered. They do this by distinguishing between the inequality constraints that are satisfied exactly and those that are not. A constraint of the form $a^T x \leq \beta$ is said to be *active* if $a^T x = \beta$, *inactive* if $a^T x < \beta$, and *violated* if $a^T x > \beta$. The motivation for this approach is that – if the set of active constraints and relevant variables were known *a priori* – the problem reduces to a smaller equality constrained sub-problem like (1.2). The KKT system for that reduced system is considered to identify a *search direction*; and the iterate is modified by this search direction within the feasible region, until some inactive constraint becomes active. The constraint is replaced in the KKT system, and this is possible at considerably less effort ($\mathcal{O}(N^2)$) as opposed to re-factoring the whole system ($\mathcal{O}(N^3)$), where N is the size of matrix (1.9). One distinguishes between primal and dual active set methods.

Barring interior-point methods, quadratic programs that are in standard form (1.2) are usually solved by an extension of the famous simplex method for linear programming by Dantzig [33]. Early considerations of this method are found in [152, 117]. More recently, we note Schönherr's algorithm and an extension to piece-wise quadratic programming [125]. The simplex method is also an iterative method that runs through a series of intermediate solutions. The main idea of this approach is to keep track of the variables that have a nonzero value in some iteration. Generally speaking, one can even allow arbitrary upper and lower bounds in place of the standard bounds. If a variable is nonzero (more generally, different from any of its bounds) it is called *basic*. Otherwise, a variable is called *non-basic*. The KKT system (1.9) is reduced to consider only the basic variables. While going through successive iterations, variables are entered and removed from the current basis. The rule by which those variables are chosen is known as (simplex) pivot rule. Even though the simplex method proved to be efficient in practice, it can lead to exponential-time behavior on certain constructed problems [82]. This is true for almost all variations of pivoting rules known up to date. It is still a major open question whether there exists a pivot rule that leads to polynomial bounds.

A common practice to make the simplex method applicable to linear and quadratic programs having inequality constraints is to add *slack variables* (see for example [26]). An inequality constraint $a^T x \leq \beta$ is transformed into an equality constraint by adding the nonnegative variable s, such that $a^T x + s = \beta$. Of course, if there are a lot of inequality constraints ($m \gg n$), this invariably leads to a blow up in the number of variables that have to be considered for the intermediate solution.

In order to retain the favorable setup for the algorithm when $m \gg n$, Schönherr combines the simplex approach with the properties of a primal active-set method, namely that a large number of slack variables does not slow down the algorithm unduly. This is achieved by considering the active constraints only. The KKT system – which is called *basis matrix* in the context of the simplex algorithm – is reduced to the relevant variables as well as to a set of active constraints. We arrive at an algorithm that performs well when $\min\{m, n\}$ is small, as we have mentioned earlier.

Last, let us point out that the solution methods described above are theoretically inferior to the *ellipsoid method* for solving convex optimization problems. This method can solve convex quadratic programming problems in weakly polynomial time [85]. This celebrated result was initially proved in 1979 by Khachiyan for linear programming [80]. In practice, however, interior-point and simplex methods prove to be much more successful. For a short primer about algorithm complexity – in particular what weak polynomiality means – see Appendix A.2.

1.1.4. Integral Factorization

There is another issue that has to be considered; that is the one of numerical accuracy. Obviously, LP and QP solvers should always compute the correct result, and even more so, they should not crash because of numerical singularities. Paraphrasing Schönherr [128] and Gärtner [56], the two extreme approaches a solver can take to address this problem are to either *expect the worst* or otherwise *hope for the best*. The former means that all computations are performed using exact arithmetic by employing an exact number type. Of course, this imposes a (possibly severe) performance penalty. It is always correct, but also always slow. The latter approach is to do all operations using standard floating point arithmetic and hoping that no instabilities arise. This is always fast and usually correct. In CGAL's QP solver a mixed strategy is employed. The original representation of the basis matrix is kept in exact arithmetic, but when it comes to deciding which variable is to enter the basis – a process that is known as *pricing* – this exact representation is converted to a floating point representation. Safety bounds on the computations with this inexact matrix are derived by Gärtner [56]. Using these bounds, it is possible to know in which cases one has to revert to exact arithmetic. This approach has the advantage of being always correct and usually fast. The fact remains, however, that the basis matrix (or more specifically its inverse) has to be computed and kept in exact arithmetic.

One of the main goals of this thesis is to expand the possibilities of CGAL's QP solver to a wider range of the parameters m and n. To do that, it becomes necessary to take advantage of the sparse structure of some inputs. This poses a limitation to the current implementation, because the basis matrix is explicitly kept as its inverse. In general, the inverse of a sparse matrix is dense. Therefore, we need another method. We will use LU factorization to obtain a sparse "inverse". Once the LU factorization of a matrix is known, it can be used to solve a linear system of equations in $\mathcal{O}(N^2)$, where N is the size of the matrix. This is in alignment with the expense that is necessary to solve the same equation system if we have the actual inverse at hand (matrix-vector multiplication). The complexity of computing either the inverse or the LU factorization from scratch is $\mathcal{O}(N^3)$. Of course, these computational expenses concern the dense case, and do not factor in the advantage we hope to gain by considering sparse systems (this will be addressed in Sections 4.4 and 5.2). First, we develop an LU procedure that works on an integral domain and does only use integral divisions. This is desirable if the input comes as elements of an integral domain such as the integers. Furthermore, we give an upper bound on the encoding size of the numbers involved in the factorization. The bound depends on the encoding size of $\det(A)$, where A is the input matrix. This is best possible in the sense that it corresponds to the same magnitude that is asserted for the final result by Cramer's rule.

Robleda conducted preliminary tests in his master's thesis [123] and was able to show that a speed-up for sparse instances seems within reach. His implementation was not free of divisions, however, and there was no bound on the size of the numbers. An interesting fact that was found by Robleda, is that if one attempts to solve the linear KKT systems by the method of *conjugate gradients* the blow-up in number size during the computation seems unmanageable. Therefore, that approach had to be abandoned.

We derive the result mentioned above independently of a similar result that has already been published in a slightly different context. The latest of a series of papers about that topic is [158] by Zhou and Jeffrey. They call the horse by a different name, and therefore we have only recently become aware of this parallel track of research. A more detailed discussion and references are found in Chapter 4, where we derive our result.

A topic that – to the best of our knowledge – has not yet been discussed in the context of integral factorizations is the one of an efficient update. This is a vital ingredient for the successful application of the CGAL QP solver to most instances. The update mirrors the changes in the basis matrix from iteration to iteration. Typically, these changes are small (constant rank updates) of the basis matrix.

Therefore, it should be possible – as it is in the case of the basis inverse – to update the factorization with less effort than it takes to do the computation from scratch. And indeed it is possible – as we show in Chapter 4 – under certain circumstances. A difficulty arises from the fact that unlike the inverse of a matrix, the LU factorization is not unique. In particular, one usually applies reordering techniques to maintain sparsity as well as possible. When performing an update on a matrix that has already been factored we have to stick to the initial ordering. This sometimes triggers a breakdown of the update procedure. Ironically, this difficulty becomes more pronounced the sparser the matrix is. We describe heuristics how the problem can be overcome, sometimes, but we still lack an adaptive reordering mechanism.

1.1.5. An Abstract View

The second major contribution of this thesis concerns a seemingly unrelated result about abstract optimization frameworks. In Chapter 2 we outline that *violator spaces* [57] exactly characterize the problems that can be solved by Clarkson's randomized algorithm[5] [27]. Violator spaces are an abstract class of optimization problems that operate on a finite ground set H, and the goal is to find a subset of $S \subseteq H$ such that S "solves" the violator space. The only basic operation allowed is the *violation test*, i.e., checking whether some $h \in H \backslash S$ violates S. If there are no violators in $H \backslash S$, we say that S is a *basis* of H and therefore solves the violator space. We develop the arguably simplest variant of Clarkson's algorithm that can still be successfully analyzed. We arrive at the essence of Clarkson's approach – unencumbered by

---

[5] Clarkson's algorithm is introduced in Section 2.1.

artificial tools that had previously been employed in the algorithm to make the analysis go through.

Now, where is the connection to quadratic programming, one might ask? It lies in the fact that the simplex algorithm for quadratic and linear programming can be formulated in terms of *violation tests*. Consider an equality constrained linear program,

$$\text{(LP)} \quad \min \quad c^T x$$

$$\text{s.t.} \quad Ax = b \tag{1.10}$$

$$x \geq 0,$$

where the number of variables n is as least as large as the number of constraints m.

It is well known [26] that in the non-degenerate case the optimal solution is uniquely defined by a subset of the variables called the *basic variables* or simply *basis*. These are the variables that have a nonzero value in the solution. In fact, let us consider the index sets $B, N \in [n]$ with $|B| = m$, $|N| = n - m$, and $B \dot{\cup} N = [n]$. Using these index sets, we can select the appropriate entries from x, c, and A. For example, A_B consists only of the columns of A whose indices are contained in B. If A_B^{-1} exists, we can derive the following formula for the values of x_B and the objective function z in terms of x_N,

$$x_B = A_B^{-1}b - A_B^{-1}A_N x_N, \tag{1.11}$$

$$z = c_B^T A_B^{-1} b + (c_N^T - c_B^T A_B^{-1} A_N) x_N. \tag{1.12}$$

If an assignment of 0 to all variables in x_N yields a nonnegative solution for x_B, we say that the variables indexed by B are a *basis* of LP (note the identical terminology as in the case of violator spaces). The whole set of assignments of values to x_B and x_N is then called a *basic feasible solution.*

Assuming that we have a basic feasible solution to start with, and by considering the vector of *reduced costs*, $\gamma = c_N^T - c_B^T A_B^{-1} A_N$, we are able to identify variables from N that may improve the solution. In particular, if $\gamma_j < 0$, we may improve the value of z by increasing the value of variable j. Variable j is called the *entering variable*[6]. If $\gamma \geq 0$ we have already found the optimal solution.

We can increase the value of the entering variable j until the value of some other variable i drops to zero. Once that variable has been identified, we can replace i by j in the basis. The variable i is therefore called the *leaving variable*. The whole process of finding i is called *ratio test*.

The important realization here is that we can regard the variables as the ground set H of a violator space. The LP pricing takes the role of the violation test and ratio test amounts to a re-computation of the basis of some subset of H. These are exactly the primitives employed in Clarkson's algorithm. As soon as we will have arrived at a vector $\gamma \geq 0$ (no more violators), we will have found a basis for the whole violator space, or in other words, the optimal solution of the LP.

The same analogy also holds for quadratic programs, but the concept of a quadratic programming basis is more complicated and left to be defined in Section 3.2.

In fact, we have an even stronger definition than required for violator spaces, because the violation is quantifiable. We know that increasing the value of the entering variable x_j by α will reduce the

---

[6] Note that we say "variable j" when we really mean variable x_j. The reader may excuse that we will use this slight abuse of notation in some places throughout the whole thesis.

objective value by $-\gamma_j\alpha$. This specialization of a violator space is called *LP-type problem*, and historically had been developed before the concept of violator spaces [100].

1.2. Statement of Results

The main results of this thesis are the following:

(i) In Chapter 2 we introduce Theorem 2.17 and Theorem 2.30, which prove results about *Clarkson's algorithm*[7]. In particular, the former extends a previously known result about Clarkson's first stage to violator spaces. The latter proves a similar result for Clarkson's second stage.

(ii) In Chapter 4 we introduce and describe Algorithm 4 (diLU) and Algorithm 6 (udiLU), which can be used to compute and update the *integral LU factorization*.

(iii) Last but not least, a considerable portion of the work that was conducted for this thesis consisted of establishing an *implementation of the integral factorization methods* in the existing quadratic programming solver of CGAL. Theoretical aspects of this part are described in Chapter 3. Technical aspects are described in Chapter 5, and experimental results, finally, are presented in Chapter 6.

---

[7] Clarkson's algorithm is going to be introduced in Section 2.1, where we also explain the meaning of the "first" and "second" stage.

Oh, many a shaft at random sent
Finds mark the archer little meant!
And many a word at random spoken
May soothe, or wound, a heart that's broken!

Sir Walter Scott

2

Violator Spaces[8]

In this chapter we are going to describe simplifications of and theoretic results about Clarkson's randomized algorithm, which is the generic tool for solving violator spaces. Note that, in this chapter, we are going to adopt a slightly different nomenclature from the one that we used in the introduction and the rest of the text. We consistently used to call the number of constraints m and the number of variables of a quadratic program n. Because the term of *combinatorial dimension* – which we will introduce later in this chapter – naturally suggests d as

---

[8] The contents of this chapter have already been published in Computational Geometry journal of Elsevier [22]. According to the publishers copyright policies reprint and archiving are permitted. We only make slight modifications in structure and content for better integration with this thesis.

a variable name, we let this override the previous convention. It will always be indicated, however, which variable denotes which quantity.

2.1. Introduction

Clarkson's algorithm. Clarkson's randomized algorithm [27] is the earliest practical linear-time algorithm for linear programming with a fixed number of variables. Combined with a later algorithm by Matoušek, Sharir and Welzl [100], it yields the best (expected) worst-case bound in the unit cost model that is known today. The combined algorithm can solve any linear program with d variables and n constraints with an expected number of $\mathcal{O}(d^2 n + \exp(\mathcal{O}(\sqrt{d \log d})))$ arithmetic operations [59].

Clarkson's algorithm consists of two primary stages, and it requires as a third stage an algorithm for solving small linear programs with $\mathcal{O}(d^2)$ constraints. The first two stages are purely combinatorial and use little problem-specific structure. One consequence of this fact is the the algorithm smoothly extends to the larger class of *LP-type problems* [100]. The bound on the running time is the same as above, for concrete problems in this class, like finding the smallest enclosing ball of a set of n points in dimension d [59].

Both primary stages of Clarkson's algorithm are based on random sampling and are conceptually simple. The main idea behind the use of randomness is that we can solve a sub-problem subject to only a small number of (randomly chosen) constraints, but still have only few (of all) constraints that are violated by the solution of the sub-problem. However, some extra machinery was originally needed to make the analysis go through. More precisely, in both stages there needed to be a check that the each individual random choice was good in a certain sense. Then in the analysis one needed to make the argument that the bad cases do not occur too often. For the first stage it was already shown by Gärtner and Welzl that these extra checks can be removed [60]. The result is what we call the *German algorithm* below. In this chapter we do the removal also for the second stage, resulting in the *Swiss algorithm*. (The names come from certain aspects of German and Swiss mentality that are reflected in the respective algorithms.) We believe that the German and the Swiss algorithm together represent the essence of Clarkson's approach.

Violator spaces. Gärtner, Matoušek, Rüst, and Škovroň proved that Clarkson's original algorithm is applicable in a still broader setting than that of LP-type problems: It works for the class of *violator spaces* [57]. At first glance, this seems to be yet another generalization to yet another abstract problem class, but as Škovroň has shown, it stops here: The class of violator spaces is the most general one for which Clarkson's algorithm is still guaranteed to work [132]. In a nutshell, the difference between LP-type problems and violator spaces is that, for the latter, the following trivial algorithm may cycle even in the non-degenerate case: maintain the optimal solution subject to a subset B of the constraints; as long as there is some constraint h that is violated by this solution, replace the current solution by the optimal solution subject to $B \cup \{h\}$, and repeat. Examples of such *cyclic* violator spaces can be found in [132]. For a easy and intuitive example see also [57].

It was unknown whether the analysis of the German algorithm (the stripped-down version of Clarkson's first stage) also works for violator spaces. For LP-type problems the analysis is nontrivial and constructs a composite LP-type problem. Here we show that this can still be done for violator spaces in essentially the same way. For the Swiss algorithm (the stripped-down version of Clarkson's second stage), we provide the first analysis at all. The fact that it works in the fully general setting of violator spaces comes naturally.

The main difference of the German and the Swiss algorithm compared to their original formulations is the following. In both stages, at some point, Clarkson's algorithm checks how many violated constraints some random sample of constraints produces. If there are too many, then the algorithm discards the sample and re-samples. The reason for this is that the analysis requires a bound on the number of violators in each step. We essentially show that this bound only needs to hold in expectation (and does so) for the analysis to go through. The checks that we mentioned before are only an analytic tool and not necessary for the algorithms to work.

Let us point out that no sub-exponential algorithm for finding the basis (that is the "solution") of a violator space is known. Therefore, we can only employ brute force to "solve" small violator spaces. Note that, e.g., in the context of linear programming, finding a basis means identifying the constraints which are tight at an optimal point. We call this the *Brute Force Algorithm* (BFA). Hence, the resulting best worst-case bound known degrades to $\mathcal{O}(d^2 n + f(d))$, where f is some exponential function of d. In this paper, we will not investigate this point further and use BFA as a black box.

The German Algorithm (GA).   Let us explain the algorithm for the problem of finding the smallest enclosing ball of a set of n points in $\mathbb{R}^d$ (this problem fits into the violator space framework). The algorithm proceeds in rounds and maintains a working set G, initialized with a subset R of r points drawn at random. In each round the smallest enclosing ball of G is being computed (by some other algorithm). For the next round the points that are unhappy with this ball (the ones that are outside) are added to G. The algorithm terminates as soon as all points are happy with the smallest enclosing ball of G.

The crucial fact that we reprove below in the violator space framework is this: The number of rounds is at most $d+2$, and for $r \approx d\sqrt{n}$, the expected maximum size of G is bounded by $\mathcal{O}(d\sqrt{n})$. This means that GA reduces a problem of size n to $d+2$ problems of expected size $\mathcal{O}(d\sqrt{n})$. We call this the German algorithm, because it takes – typically German – one decision in the beginning which is then efficiently pulled through.

The Swiss Algorithm (SA). Like GA, this algorithm proceeds in rounds, but it maintains a voting box that initially contains one slip per point. In each round a set of r slips is drawn at random from the voting box, and the smallest enclosing ball of the corresponding set R is computed (by some other algorithm). For the next round all slips are put back, and on top of that, the number of slips of the unhappy points is doubled. The algorithm terminates as soon as all points are happy with the smallest enclosing ball of the sample R.

Below we will prove the following: If $r \approx d^2$, the expected number of rounds is $\mathcal{O}(\log n)$. This means that SA reduces a problem of size n to $\mathcal{O}(\log n)$ problems of size $\mathcal{O}(d^2)$. We call this the Swiss algorithm, because it takes – typically Swiss – many independent local decisions that magically fit together in the end.

Hypercube partitions. A hypercube partition is a partition of the vertices of the hypercube such that every element of the partition is the set of vertices of some sub-cube. It was known that every *non-degenerate* violator space induces a hypercube partition [101, 99]. We prove here that also the converse is true, meaning that we obtain an alternative characterization of the class of violator spaces. While this result is not hard to obtain, it may be useful in the future for the problem of counting violator spaces. The initial bounds provided by Škovroň are still the best known ones [132].

Applications.   We would love to present a number of convincing applications of the violator space framework, and in particular of the German and the Swiss algorithm for violator spaces. Unfortunately, we cannot. There is one known application of Clarkson's algorithm that really requires it to work for violator spaces and not just LP-type problems [57]; this application (solving generalized P-matrix linear complementarity problems with a fixed number of blocks) benefits from our improvements in the sense that now also the German and the Swiss algorithm are applicable to it (with less random resources than Clarkson's algorithm).

Our main contributions are therefore theoretical: We show that Clarkson's second stage can be simplified (resulting in the Swiss algorithm), and this result is new even for LP-type problems and linear programming. The fact that Clarkson's first stage can be simplified (resulting in the German algorithm) was known for LP-type problems; we extend it to violator spaces, allowing the German algorithm to be used for solving generalized P-matrix linear complementarity problems with a fixed number of blocks.

We believe that our version of Clarkson's algorithm is the most simple variant that can still successfully be analyzed.

2.2. Prerequisites

2.2.1. The Sampling Lemma

The following lemma is due to Gärtner and Welzl in [60] and was adapted to violator spaces in [57]. We repeat it here for the sake of completeness, and because its proof and formulation are concise. Let S be a set of size n, and $\varphi : 2^S \to \mathbb{R}$ a function that maps any set $R \subseteq S$ to some value $\varphi(R)$. Define

$$\mathsf{V}(R) := \{\, s \in S \backslash R \mid \varphi(R \cup \{s\}) \neq \varphi(R)\,\}, \qquad (2.1)$$

$$\mathsf{X}(R) := \{\, s \in R \mid \varphi(R \backslash \{s\}) \neq \varphi(R)\,\}. \qquad (2.2)$$

$\mathsf{V}(R)$ is the set of *violators* of R, while $\mathsf{X}(R)$ is the set of *extreme elements* in R. Obviously,

$$s \text{ violates } R \Leftrightarrow s \text{ is extreme in } R \cup \{s\}.$$

For a random sample R of size r, i.e., a set R chosen uniformly at random from the set $\binom{S}{r}$ of all r-element subsets of S, we define random variables $\mathsf{V}_r : R \mapsto |\mathsf{V}(R)|$ and $\mathsf{X}_r : R \mapsto |\mathsf{X}(R)|$, and we consider the expected values

$$v_r := \mathrm{E}[\mathsf{V}_r],$$

$$x_r := \mathrm{E}[\mathsf{X}_r].$$

Lemma 2.3 (Sampling Lemma, [60, 57]). *For* $0 \leq r < n$,

$$\frac{v_r}{n-r} = \frac{x_{r+1}}{r+1}.$$

Proof. Using the definitions of v_r and x_{r+1} as well as (2.2.1), we can argue as follows,

$$\binom{n}{r} v_r = \sum_{R \in \binom{S}{r}} \sum_{s \in S \setminus R} [s \text{ violates } R]$$

$$= \sum_{R \in \binom{S}{r}} \sum_{s \in S \setminus R} [s \text{ is extreme in } R \cup \{s\}]$$

$$= \sum_{Q \in \binom{S}{r+1}} \sum_{s \in Q} [s \text{ is extreme in } Q]$$

$$= \binom{n}{r+1} x_{r+1}.$$

Here, $[\cdot]$ is the indicator variable for the event in brackets. Finally, $\binom{n}{r+1} / \binom{n}{r} = (n-r)/(r+1)$. $\qquad\square$

2.2.2. Definition of Violator Spaces

Definition 2.4. *A* violator space *is a pair* (H, V), *where* H *is a finite set and* V *is a mapping* $2^H \to 2^H$ *such that the following two conditions are fulfilled.*

Consistency: $G \cap \mathsf{V}(G) = \emptyset$ *holds for all* $G \subseteq H$, *and*
Locality: *for all* $F \subseteq G \subseteq H$, *where* $G \cap \mathsf{V}(F) = \emptyset$, *we have* $\mathsf{V}(G) = \mathsf{V}(F)$.

Lemma 2.5 (Lemma 17, [57]). *Any violator space* (H, V) *satisfies* monotonicity *defined as follows:*

Monotonicity: $\mathsf{V}(F) = \mathsf{V}(G)$ *implies* $\mathsf{V}(E) = \mathsf{V}(F) = \mathsf{V}(G)$ *for all sets* $F \subseteq E \subseteq G \subseteq H$.

Proof. Assume that $\mathsf{V}(E) \neq \mathsf{V}(F), \mathsf{V}(G)$. Then locality yields $\emptyset \neq E \cap \mathsf{V}(F) = E \cap \mathsf{V}(G)$ which contradicts consistency. $\qquad\square$

Definition 2.6. *Consider a violator space* (H, V).

(i) *We say that* $B \subseteq H$ *is a* basis *if for all proper subsets* $F \subset B$ *we have* $B \cap \mathsf{V}(F) \neq \emptyset$. *For* $G \subseteq H$, *a* basis of G *is a minimal subset* B *of* G *with* $\mathsf{V}(B) = \mathsf{V}(G)$. *A* basis in (H, V) *is a basis of some set* $G \subseteq H$.

(ii) *The* combinatorial dimension *of* (H, V), *denoted by* $\dim(H, \mathsf{V})$, *is the size of the largest basis in* (H, V).

(iii) (H, V) *is* non-degenerate *if every set* $G \subseteq H$, $|G| \geq \dim(H, \mathsf{V})$, *has a unique basis. Otherwise* (H, V) *is* degenerate.

Observe that a minimal subset $B \subseteq G$ with $\mathsf{V}(B) = \mathsf{V}(G)$ is indeed a basis: Assume for contradiction that there is a set $F \subset B$ such that $B \cap \mathsf{V}(F) = \emptyset$. Locality then yields $\mathsf{V}(B) = \mathsf{V}(F) = \mathsf{V}(G)$, which contradicts the minimality of B. Also, note that, because of consistency, any basis B of H has no violators $\mathsf{V}(H) = \mathsf{V}(B) = \emptyset$.

Corollary 2.7 (of Lemma 2.3). *Let* (H, V) *be a violator space of combinatorial dimension* d, *and* $|H| = n$. *If we choose a subset* $R \subseteq H$, $|R| = r \leq n$, *uniformly at random, then*

$$\mathrm{E}[|\mathsf{V}(R)|] \leq d\frac{n-r}{r+1}.$$

Proof. The corollary follows from the Sampling Lemma 2.3, with the observation that $|X(R)| \leq d$, $\forall R \subseteq H$. □

2.3. Clarkson's Algorithm Revisited

Clarkson's algorithm can be used to compute a basis of some violator space (H, V), $n = |H|$. It consists of two separate stages and the

Brute Force Algorithm (BFA). The results about the running time and the size of the sets involved are summarized in Theorem 2.17 and Theorem 2.30.

The main idea of both stages (GA and SA) is the following: We draw a random sample $R \subseteq H$ of size $r = |R|$ and then compute a basis of R using some other algorithm. The crucial point here is that $r \ll n$ hopefully. Obviously, such an approach may fail to find a basis of H, and we might have to reconsider and enter a second round. That is the point at which GA and SA most significantly differ.

In both stages we assume that the size of the ground set n is larger than r, such that we can actually draw a sample of that size. We can assume this w.l.o.g., because it is easy to incorporate an if statement at the beginning that directly calls the other algorithm should n be too small.

2.3.1. The German Algorithm (GA)

This algorithm works as follows. Let (H, V) be a violator space, $|H| = n$, and $\dim(H, \mathsf{V}) = d$. We draw a random sample $R \subseteq H$, $r = d\sqrt{n/2}$, only once, and initialize our working set G with R. Then we enter a repeat loop, in which we compute a basis B of G and check whether there are any violators in H. If no, then we are done and return the basis B. If yes, then we add those violators to our working set G and repeat the procedure.

The analysis will show that (i) the number of rounds is bounded by $d + 1$, and (ii) the size of G in any round is bounded by $\mathcal{O}(d\sqrt{n})$.

Algorithm 1: German Algorithm (GA)

Input : Violator space (H, V), with $|H| = n$ and $\dim(H, \mathsf{V}) = d$
Output: A basis B of (H, V)

1 $r \leftarrow d\sqrt{n/2}$;
2 Choose $R \subseteq H$ u.a.r., with $|R| = r$;
3 $G \leftarrow R$;
4 **repeat**
5 | $B \leftarrow \mathsf{SA}(G, \mathsf{V}|_G)$;
6 | $G \leftarrow G \cup \mathsf{V}(B)$;
7 **until** $\mathsf{V}(B) = \emptyset$;
8 **return** B

We will adopt some useful notations which we will use in the following proofs. First, let us point out that the notation $\mathsf{V}|_F$ refers to the violator mapping restricted to some set $F \subseteq H$.

Definition 2.8. *For $i \geq 0$, by*

$$B_R^{(i)}, \ V_R^{(i)}, \ and \ G_R^{(i)}$$

we denote the sets B, $\mathsf{V}(B)$, and G computed in round i of the repeat loop above. Furthermore, we set $G_R^{(0)} := R$, while $B_R^{(0)}$ and $V_R^{(0)}$ are undefined. In particular, we have that $B_R^{(i)}$ is a basis of $G_R^{(i-1)}$, and $V_R^{(i)} = \mathsf{V}(G_R^{(i-1)})$. If the algorithm performs exactly ℓ rounds, sets with indices $i > \ell$ are defined to be the corresponding sets of round ℓ.

The next one is an auxiliary lemma that we will need further on in the analysis. It is a generalization of the fact that there is at least one element of the basis of H found as a violator in every round (see also Lemma 2.21).

Lemma 2.9. *For $j < i \leq \ell$, $B_R^{(i)} \cap V_R^{(j)} \neq \emptyset$.*

Proof. Assume that $B_R^{(i)} \cap V_R^{(j)} = \emptyset$. Together with consistency, $G_R^{(j-1)} \cap V_R^{(j)} = \emptyset$, this implies

$$(B_R^{(i)} \cup G_R^{(j-1)}) \cap V_R^{(j)} = \emptyset.$$

Now, applying locality and the definition of basis, we get

$$\mathsf{V}(B_R^{(i)} \cup G_R^{(j-1)}) = V_R^{(j)} = \mathsf{V}(B_R^{(j)}). \tag{2.10}$$

On the other hand, since $V_R^{(i)} = \mathsf{V}(B_R^{(i)})$ and $B_R^{(i)} \subseteq B_R^{(i)} \cup G_R^{(j-1)} \subseteq G_R^{(i-1)}$, we can apply monotonicity and derive

$$V_R^{(i)} = \mathsf{V}(B_R^{(i)}) = \mathsf{V}(B_R^{(i)} \cup G_R^{(j-1)}). \tag{2.11}$$

Note that $V(B_R^{(j)}) \subseteq G_R^{(i-1)}$, because G always contains the violators from previous rounds. Additionally, by equations (2.10) and (2.11) we have that $V_R^{(i)} = \mathsf{V}(B_R^{(i)} \cup G_R^{(j-1)}) = \mathsf{V}(B_R^{(j)})$. Thus, we can build a contradiction of consistency,

$$G_R^{(i-1)} \cap V_R^{(i)} \supseteq \mathsf{V}(B_R^{(j)}) \cap V_R^{(i)} = \mathsf{V}(B_R^{(j)}) \neq \emptyset.$$

The last inequality holds because j is not the last round. $\square$

The following lemma is the crucial result that lets us interpret the development of the set G in the German Algorithm (Algorithm 1) as a violator space itself.

Lemma 2.12. *Let (H, V) be a violator space of combinatorial dimension d. For any subset $R \subseteq H$ define*

$$\Gamma(R) := (V_R^{(1)}, \ldots, V_R^{(d)}). \tag{2.13}$$

Using this we can define a new violator mapping as follows,

$$\mathsf{V}'(R) := \{\, h \in H \backslash R \mid \Gamma(R) \neq \Gamma(R \cup \{h\}) \,\}. \tag{2.14}$$

Then the following statements are true:

(i) (H, V') *is a violator space of combinatorial dimension at most* $\binom{d+1}{2}$.

(ii) *The set* $\mathsf{V}'(R)$ *is given by*

$$\mathsf{V}'(R) = V_R^{(1)} \cup \ldots \cup V_R^{(d)} = G_R^{(d)} \backslash R.$$

(iii) *If* (H, V) *is non-degenerate, then so is* (H, V').

To prove Lemma 2.12 we first need an auxiliary claim. Note that the symbol $\dot\cup$ denotes disjoint union.

Claim 2.15. *Let* Q *be any set with* $Q = R \mathbin{\dot\cup} T \subseteq H$ *and* $i < d$. *If*

$$V_Q^{(j+1)} = V_R^{(j+1)}, \qquad j \le i,$$

then

$$G_Q^{(j)} = G_R^{(j)} \mathbin{\dot\cup} T, \qquad j \le i+1.$$

Proof of Claim 2.15. We prove the claim by induction on i. First, if $i = 0$ the precondition reads $\mathsf{V}(Q) = \mathsf{V}(R)$. It follows that $G_Q^{(1)} = Q \cup \mathsf{V}(Q) = (R \mathbin{\dot\cup} T) \cup \mathsf{V}(R) = G_R^{(1)} \mathbin{\dot\cup} T$.

Suppose the claim is true for $j \le i$. From $V_Q^{(i+1)} = V_R^{(i+1)}$ we can deduce

$$G_Q^{(i+1)} = G_Q^{(i)} \cup V_Q^{(i+1)} = (G_R^{(i)} \mathbin{\dot\cup} T) \cup V_R^{(i+1)} = G_R^{(i+1)} \mathbin{\dot\cup} T.$$

$\square$

Before we proceed to the proof of Lemma 2.12 let us first state the consequences, which we obtain by applying Lemma 2.3 to the violator space that we constructed.

Theorem 2.16 (Theorem 5.5 of [60]). *For $R \subseteq H$ with $|H| = n$, and a random sample of size r,*

$$\mathrm{E}[|G_R^{(d)}|] \leq \binom{d+1}{2} \frac{n-r}{r+1} + r.$$

Choosing $r = d\sqrt{n/2}$ yields

$$\mathrm{E}[|G_R^{(d)}|] \leq 2(d+1)\sqrt{\frac{n}{2}}.$$

Proof of Theorem 2.16. The first inequality directly follows from the sampling lemma (Lemma 2.3), applied to the violator space (H, V'), together with part (ii) of Lemma 2.12. The second inequality follows from plugging in the value for r.   □

Let us now come back to the previous lemma.

Proof of Lemma 2.12.
Proof of (i). We first need to check consistency and locality as defined in Definition 2.4.

Consistency is easy, by the definition of V'. Since the violators of $R \subseteq H$ are chosen from $H \backslash R$ exclusively, we can be sure that $R \cap \mathsf{V}'(R) = \emptyset$ for all R.

Let us recall what locality means. For sets $R \subseteq Q \subseteq H$, if $Q \cap \mathsf{V}'(R) = \emptyset$, then $\mathsf{V}'(Q) = \mathsf{V}'(R)$. This we are going to prove by induction on the size of $Q \backslash R$. If $|Q \backslash R| = 0$, then the two sets are the same, and locality is obviously fulfilled. Now, suppose that $|Q \backslash R| = i$ and locality is true for any smaller value $j < i$. Consider some set S fulfilling $R \subseteq S \subset Q$ and $Q = S \,\dot{\cup}\, \{q\}$. First note that, if $Q \cap V'(R) = \emptyset$, then also $S \cap V'(R) = \emptyset$. Therefore, the precondition for the induction hypothesis is fulfilled, and we can conclude that $V'(R) = V'(S)$. Bearing this in mind, we can make the following derivation,

$$
\begin{aligned}
Q \cap V'(R) = \emptyset \;\Rightarrow\quad & Q \cap V'(S) = \emptyset \\
\overset{q \in Q}{\Rightarrow}\quad & q \notin V'(S) \\
\overset{\text{Def. (2.1)}}{\Rightarrow}\quad & \Gamma(S) = \Gamma(S \,\dot{\cup}\, \{q\}) = \Gamma(Q) \\
\overset{\text{Def. (2.1)}}{\Rightarrow}\quad & V'(S) = V'(Q) \\
\Rightarrow\quad & V'(R) = V'(Q).
\end{aligned}
$$

That shows the locality of the violator space (H, V').

We still have to show that (H, V') has combinatorial dimension at most $\binom{d+1}{2}$. To this end we prove that $V'(B_R) = V'(R)$, where

$$
B_R := R \cap \bigcup_{i=1}^{d} B_R^{(i)}.
$$

Note that B_R, as we will show in (iii), is in fact the unique basis of the set $R \subseteq H$. By bounding the size of B_R we therefore bound the combinatorial dimension of (H, V'). Equivalent to $V'(B_R) = V'(R)$ we show that $V_{B_R}^{(j)} = V_R^{(j)}$, for $1 \leq j \leq d$, using induction on j. For $j = 1$ we get

$$
V(R) = V(B_R \cup (R \backslash B_R)) = V(B_R),
$$

because $R \backslash B_R$ is disjoint from $B_R^{(1)}$, the basis of R. Therefore, $R \backslash B_R = R \backslash \bigcup_{i=1}^{d} B_R^{(i)}$ can be removed from R without changing the set of violators.

Now assume that the statement holds for $j \leq d - 1$ and consider the case $j = d$. By Claim 2.15, we get $G_R^{(j-1)} = G_{B_R}^{(j-1)} \;\dot\cup\; (R \backslash B_R)$. Since $R \backslash B_R$ is disjoint from the basis $B_R^{(j)}$ of $G_R^{(j-1)}$ it follows that

$$V_R^{(j)} = \mathsf{V}(G_R^{(j-1)}) = \mathsf{V}(G_{B_R}^{(j-1)} \;\dot\cup\; (R \backslash B_R)) = \mathsf{V}(G_{B_R}^{(j-1)}) = V_{B_R}^{(j)}.$$

To bound the size of B_R, we observe that

$$|R \cap B_R^{(i)}| \le d + 1 - i,$$

for all $i \le \ell$ (the number of rounds in which $\mathsf{V}(B) \ne \emptyset$). This follows from Lemma 2.9. $B_R^{(i)}$ has at least one element in each of the $i - 1$ sets $V_R^1, \ldots, V_R^{(i-1)}$, which are in turn disjoint from R. Hence we get

$$|B_R| \le \sum_{i=1}^{\ell} |R \cap B_R^{(i)}| \le \binom{d+1}{2}.$$

Proof of (ii). We show that if some constraint $q \in H$ is in $\mathsf{V}'(R)$ then it is also in $V_R^{(i)}$ for some $1 \le i \le d$. On the other hand if $q \notin \mathsf{V}'(R)$ then q is not in any of the $V_R^{(i)}$, $1 \le i \le d$. This proves the statement of (ii).

Assume $q \in \mathsf{V}'(R)$ and let $Q := R \cup \{q\}$. Consider the largest index $i < d - 1$, such that

$$V_R^{(j+1)} = V_Q^{(j+1)}, \quad j \le i.$$

Note that such an index i must exist, because $\mathsf{V}'(R) \ne \mathsf{V}'(Q)$, which simply follows from $q \in \mathsf{V}'(R)$ and $q \notin \mathsf{V}'(Q)$. Then, from Claim 2.15 it follows that $G_Q^{(i+1)} = G_R^{(i+1)} \;\dot\cup\; \{q\}$, and by assumption on i we know that $V_R^{(i+2)} \ne V_Q^{(i+2)}$. Therefore, by the contrapositive of locality, we conclude $(G_R^{(i+1)} \;\dot\cup\; \{q\}) \cap \mathsf{V}(G_R^{(i+1)}) \ne \emptyset$. This means that

$q \in \mathsf{V}(G_R^{(i+1)}) = V_R^{(i+2)}$, because otherwise the consistency of $G_R^{(i+1)}$ would be violated.

On the other hand, if $q \notin \mathsf{V}'(R)$, then $\mathsf{V}'(R) = \mathsf{V}'(Q)$, or equivalently $V_R^{(i)} = V_Q^{(i)}$, for $1 \leq i \leq d$. However, because (H, V) is consistent it follows that $q \notin V_Q^{(i)}$, and therefore $q \notin V_R^{(i)}$, for $1 \leq i \leq d$.

Proof of (iii). Non-degeneracy of (H, V') follows if we can show that every set $R \subseteq H$ has the set B_R as its unique basis. To this end we prove that whenever we have $L \subseteq R$ with $\mathsf{V}'(L) = \mathsf{V}'(R)$, then $B_R \subseteq L$.

Fix $L \subseteq R$ with $\mathsf{V}'(L) = \mathsf{V}'(R)$, i.e.,

$$V_R^{(i)} = V_L^{(i)}, \ 1 \leq i \leq d.$$

Claim 2.15 then implies

$$G_R^{(i)} = G_L^{(i)} \ \dot{\cup} \ (R \backslash L), \ 0 \leq i \leq d,$$

and the non-degeneracy of (H, V) yields that $G_R^{(i)}$ and $G_L^{(i)}$ have the same unique basis $B_R^{(i+1)}$, for all $0 \leq i \leq d$. Note that $B_R^{(i+1)}$ is indeed contained in $G_L^{(i)}$, because $\mathsf{V}(G_L^{(i)}) = \mathsf{V}(G_R^{(i)}) = \mathsf{V}(G_L^{(i)} \dot{\cup} (R \backslash L)) = \mathsf{V}(B_R^{(i+1)})$ for $0 \leq i \leq d$. That means, if there exists a basis of $G_L^{(i)}$, which by definition is also a basis of $G_R^{(i)}$, but distinct from $B_R^{(i+1)}$, non-degeneracy is violated.

It follows that $G_L^{(d-1)}$ contains

$$\bigcup_{i=1}^{d} B_R^{(i)},$$

so L contains

$$L \cap \bigcup_{i=1}^{d} B_R^{(i)} = R \cap \bigcup_{i=1}^{d} B_R^{(i)}.$$

The latter equality holds because $R \backslash L$ is disjoint from $G_L^{(d)}$, thus in particular from the union of the $B_R^{(i)}$.   □

Theorem 2.17. *Let (H, V) be a violator space of combinatorial dimension d, and $n = |H|$. Then the algorithm GA computes a basis of (H, V) with at most $d + 1$ calls to SA, with an expected number of at most $\mathcal{O}(d\sqrt{n})$ constraints each.*

Proof. According to Lemma 2.9 (and maybe more intuitively according to Lemma 2.21), in every round except the last one we add at least one element of any basis of (H, V) to G. Since the size of the basis is bounded by d we get that the number of rounds is at most $d + 1$. Furthermore, according to Theorem 2.16, and our choice $r = d\sqrt{n/2}$, the expected size of G will not exceed $2(d+1)\sqrt{n/2}$ in any round.   □

2.3.2. The Swiss Algorithm (SA)

The algorithm SA proceeds similar as the first one. Let the input be a violator space (H, V), $|H| = n$, and $\dim(H, \mathsf{V}) = d$.

First, let us (re)introduce the notation $R^{(i)}$, $B^{(i)}$, and $V^{(i)}$ for $i \geq 1$, similar as in Definition 2.8, for the sets R, B and $\mathsf{V}(R)$ of round i respectively. The set $B^{(i)}$ is a basis of $R^{(i)}$ and $V^{(i)} = \mathsf{V}(R^{(i)}) = \mathsf{V}(B^{(i)})$. Since we draw a random sample in every round it does not make sense to index the sets $B^{(i)}$ and $V^{(i)}$ by R, so we drop this subscript.

After the initialization, we enter the first round and choose a random sample $R^{(1)}$ of size $r = 2d^2$ uniformly at random from H. Then we compute an intermediate basis $B^{(1)}$ of the violator space

$(R^{(1)}, \mathsf{V}|_{R^{(1)}})$ by using BFA as a black box. In the next step, we compute the set of violated constraints $V^{(1)}$. So far, it is the same thing as the first stage. But now, instead of enforcing the violated constraints by adding them to the active set, we increase the probability that the violated constraints are chosen in the next round. This is achieved by means of the multiplicity or weight variable μ.

Algorithm 2: Swiss Algorithm (SA)

Input	: Violator space (H, V), $|H| = n$, and $\dim(H, \mathsf{V}) = d$
Output	: A basis B of (H, V)

1    $\mu_h \leftarrow 1$ for all $h \in H$;

2    $r \leftarrow 2d^2$;

3   **repeat**

4   choose random R from H according to μ;

5       $B \leftarrow \mathsf{BFA}(R, \mathsf{V}|_R)$;

6       $\mu_h \leftarrow 2\mu_h$ for all $h \in \mathsf{V}(B)$;

7   **until** $V(B) = \emptyset$;

8   **return** B

Definition 2.18. *With every $h \in H$ we associate the* multiplicity $\mu_h \in \mathbb{N}$. *For an arbitrary set $F \subseteq H$ we define the* cumulative multiplicity *as*

$$\mu(F) := \sum_{h \in F} \mu_h.$$

For the analysis we also need to keep track of this value across different iterations of the algorithm. For $i \geq 0$ we will use $\mu_h^{(i)}$ (and $\mu^{(i)}(F)$) to denote the (cumulative) multiplicity at the end of round i. We define $\mu_h^{(0)} := 1$ for any $h \in H$, and therefore $\mu^{(0)}(F) = |F|$.

Now back to the algorithm. To increase the probability that a constraint $h \in V^{(i)}$ is chosen in the random sample of round $i + 1$ we double the multiplicity of h, i.e., $\mu_h^{(i)} := 2\mu_h^{(i-1)}$.

The multiplicities determine how the random sample $R^{(i+1)}$ is chosen. To this end we construct a multiset $\hat{H}^{(i+1)}$ to which we add $\mu_h^{(i)}$ copies of every element $h \in H$. To simplify notation, let us for a moment fix the round $i+1$ and drop the corresponding superscript.

We define the function $\phi : 2^H \to 2^{\hat{H}}$ as the function that maps a set of elements from H to the set of corresponding elements in $\hat{H}$, i.e., for $F \subseteq H$,

$$\phi(F) := \bigcup_{h \in F} \{h_1, \ldots, h_{\mu_h}\}, \tag{2.19}$$

where the h_j, $1 \le j \le \mu_h$, are the distinct copies of h. For example, $\hat{H} = \phi(H)$. Conversely, let $\psi : 2^{\hat{H}} \to 2^H$ be the function that collapses a given subset of $\hat{H}$ to their original elements in H, i.e., for $\hat{F} \subseteq \hat{H}$,

$$\psi(\hat{F}) := \{h \in H \mid \phi(\{h\}) \cap \hat{F} \ne \emptyset\}. \tag{2.20}$$

Reintroducing the superscript $i+1$ we can simply say that we construct $\hat{H}^{(i+1)} = \phi(H)$ using the multiplicities from round i. The sample $\hat{R}^{(i+1)}$ is then chosen u.a.r. from the r-subsets of $\hat{H}^{(i+1)}$. In the following the multiset property will not be important any more and we can discard multiple entries to obtain $R^{(i+1)} = \psi(\hat{R}^{(i+1)})$. Note that $1 \le |R^{(i+1)}| \le r$. Then we continue as in round 1. Note that in the first round this is in fact equivalent to choosing an r-subset u.a.r. from H, because $\mu_h^{(0)} = 1$ for all $h \in H$.

The algorithm terminates as soon as $V^{(\ell)} = \emptyset$ for some round $\ell \ge 1$ and returns the basis $B^{(\ell)}$.

Let us first discuss an auxiliary lemma similar to Lemma 2.9.

Lemma 2.21 (Observation 22 in [57]). *Let (H, V) be a violator space, $F \subseteq G \subseteq H$, and $G \cap \mathsf{V}(F) \ne \emptyset$. Then $G \cap \mathsf{V}(F)$ contains at least one element from every basis of G.*

Proof. Since the proof is short we repeat it here. Let B be some basis of G and assume that $B \cap G \cap \mathsf{V}(F) = B \cap \mathsf{V}(F) = \emptyset$. From consistency we get $F \cap \mathsf{V}(F) = \emptyset$. Together this implies

$$(B \cup F) \cap \mathsf{V}(F) = \emptyset.$$

Applying locality and monotonicity, we get

$$\mathsf{V}(F) = \mathsf{V}(B \cup F) = \mathsf{V}(G),$$

meaning that $G \cap \mathsf{V}(G) = G \cap \mathsf{V}(F) = \emptyset$, a contradiction. □

The analysis of SA will show that the elements in any basis B of H will increase their multiplicity so quickly that they are chosen with high probability after a logarithmic number of rounds. This, of course, means that the algorithm will terminate, because there will be no violators. Formally, we will have to employ trick though. We will consider a modification of SA that runs forever, regardless of the current set of violators. Let us call the modified algorithm SA-forever. We call a particular round i *controversial* if $V^{(i)} \neq \emptyset$. Furthermore, let C_ℓ be the event that the first ℓ rounds are controversial in SA-forever.

Lemma 2.22. *Let* (H, V) *be a violator space,* $|H| = n$, $\dim (H, \mathsf{V}) = d$, B *any basis of* H, *and* $k \in \mathbb{N}$ *some positive integer. Then, in SA-forever, the following holds for the expected cumulative multiplicity of* B *after* kd *rounds,*

$$2^k \Pr[C_{kd}] \leq \mathrm{E}[\mu^{(kd)}(B)].$$

Proof. In any controversial round, Lemma 2.21 asserts that $B \cap V^{(i)} \neq \emptyset$. So, in every controversial round, the multiplicity of at least one element in B is doubled. Therefore, by conditioning on the event that the first kd rounds are controversial, there must be a constraint in B that has been doubled at least k times (recall that $|B| \leq d$). It follows that $\mathrm{E}[\mu^{(kd)}(B)] = \mathrm{E}[\mu^{(kd)}(B) \mid C_{kd}] \Pr[C_{kd}] + \mathrm{E}[\mu^{(kd)}(B) \mid \overline{C_{kd}}] \Pr[\overline{C_{kd}}] \geq 2^k \Pr[C_{kd}]$. $\hfill\square$

Lemma 2.23. *Let* (H, V) *be a violator space,* $|H| = n$, $\dim(H, \mathsf{V}) = d$, B *any basis of* H, *and* $k \in \mathbb{N}$ *some positive integer. Then, in* SA-forever, *the following holds for the expected cumulative multiplicity of* B *after* kd *rounds,*

$$\mathrm{E}[\mu^{(kd)}(B)] \leq n \left(1 + \frac{d}{r}\right)^{kd}.$$

Proof. Let us point out first, that the following analysis goes through for SA-forever as well as for SA, but to make it match Lemma 2.22 we formulated it using the former.

Note that $\mathrm{E}[\mu^{(kd)}(B)] \leq \mathrm{E}[\mu^{(kd)}(H)]$, because $B \subseteq H$. Therefore, if we show the upper bound for the latter expectation we are done. Let $\ell := kd$ be the number of rounds, and $\Delta^{(i)}(F) := \mu^{(i)}(F) - \mu^{(i-1)}(F)$ the increase of multiplicity from one round to another, for any $i \geq 1$ and $F \subseteq H$. We write the expected weight of H after ℓ rounds as the sum of the initial weight plus the expected increase in weight in every round from 1 to ℓ,

$$\mathrm{E}[\mu^{(\ell)}(H)] = \mathrm{E}[\mu^{(0)}(H)] + \sum_{i=1}^{\ell} \mathrm{E}[\Delta^{(i)}(H)]. \tag{2.24}$$

The first term is easy, $\mathrm{E}[\mu^{(0)}(H)] = n$, and the second term we write as a conditional expectation, assuming that the weight in round $i-1$

was t,

$$\sum_{i=1}^{\ell} \mathrm{E}[\Delta^{(i)}(H)] = \sum_{i=1}^{\ell} \sum_{t=0}^{\infty} \mathrm{E}[\Delta^{(i)}(H) | \mu^{(i-1)}(H) = t] \Pr[\mu^{(i-1)}(H) = t].$$
(2.25)

Now comes the crucial step. According to Lemma 2.3 we can upper bound $\mathrm{E}[\Delta^{(i)}(H) | \mu^{(i-1)}(H) = t]$ by interpreting it as the expected number of violators of a multiset extension of (H, V). To this end we construct a violator space $(\hat{H}^{(i)}, \hat{\mathsf{V}})$, where $\hat{H}^{(i)} = \phi(H)$ using the multiplicities from round $i - 1$. Let us fix round i and drop the superscript for the moment. For any $\hat{F} \subseteq \hat{H}$ we define

$$\hat{\mathsf{V}}(\hat{F}) := \phi(\mathsf{V}(\psi(\hat{F}))).$$
(2.26)

We observe that $(\hat{H}, \hat{\mathsf{V}})$ is indeed a violator space. For $\hat{F} \subseteq \hat{H}$, consistency is preserved, because from consistency of (H, V) it follows that $\phi(\psi(\hat{F})) \cap \phi(\mathsf{V}(\psi(\hat{F}))) = \emptyset$, and knowing $\hat{F} \subseteq \phi(\psi(\hat{F}))$, we can conclude consistency of $(\hat{H}, \hat{\mathsf{V}})$. Similarly, for $\hat{F} \subseteq \hat{G} \subseteq \hat{H}$, locality of (H, V) tells us that if $\phi(\psi(\hat{G})) \cap \phi(\mathsf{V}(\psi(\hat{F}))) = \emptyset$ then $\phi(\mathsf{V}(\psi(\hat{F}))) = \phi(\mathsf{V}(\psi(\hat{G})))$, and knowing $\hat{G} \subseteq \phi(\psi(\hat{G}))$, locality of $(\hat{H}, \hat{\mathsf{V}})$ follows.

The violator space we just constructed has the same ground set $\hat{H}$ by means of which we draw the random sample R in every round. By supplying a valid violator mapping we asserted that we can apply the sampling lemma to that process. Some thinking reveals that $d = \dim(H, \mathsf{V}) = \dim(\hat{H}, \hat{\mathsf{V}})$ (even though we introduced degeneracy), and we can conclude that

$$\mathrm{E}[\Delta^{(i)}(H) | \mu^{(i-1)}(H) = t] = \mathrm{E}[|\hat{\mathsf{V}}(\hat{R}^{(i)})|] \leq d \frac{t-r}{r+1}.$$
(2.27)

Therefore we get the simplified expression

$$\begin{aligned}
\mathrm{E}[\mu^{(\ell)}(H)] \;\leq\; & n + \sum_{i=1}^{\ell} \sum_{t=0}^{\infty} d\frac{t-r}{r+1} \Pr[\mu^{(i-1)}(H) = t] \\[2mm]
=\; & n + \sum_{i=1}^{\ell} \left(\frac{d}{r+1} \sum_{t=0}^{\infty} t \Pr[\mu^{(i-1)}(H) = t] \right. \\[2mm]
& \left. -\frac{dr}{r+1} \sum_{t=0}^{\infty} \Pr[\mu^{(i-1)}(H) = t] \right) \\[2mm]
=\; & n + \frac{d}{r+1} \sum_{i=1}^{\ell} \mathrm{E}[\mu^{(i-1)}(H)] - \ell\frac{dr}{r+1}.
\end{aligned}$$

The first line is derived from (2.24), (2.25), and (2.27). The rest is routine. Dropping the last term we get the following recursive equation,

$$\mathrm{E}[\mu^{(\ell)}(H)] \leq n + \frac{d}{r+1} \sum_{i=0}^{\ell-1} \mathrm{E}[\mu^{(i)}(H)],$$

which easily resolves to the claimed bound. $\square$

Using $\ell = kd$, and combining Lemma 2.22 and 2.23, we now know that

$$2^k \; \Pr[C_\ell] \leq n \left(1 + \frac{d}{r} \right)^\ell.$$

This inequality gives us a useful upper bound on $\Pr[C_\ell]$, because the left-hand side power grows faster than the right-hand side power as a function of ℓ, given that r is chosen large enough.

Let us choose $r = c\,d^2$ for some constant $c > \log_2 e \approx 1.44$. We obtain

$$\Pr[C_\ell] \leq n \left(1 + \frac{1}{c\,d} \right)^\ell / 2^k \leq n\, 2^{(\ell \log_2 e)/(c\,d) - k},$$

using $1 + x \leq e^x = 2^{x \log_2 e}$ for all x. This further gives us

$$\Pr[C_\ell] \leq n\alpha^\ell, \qquad (2.28)$$

$$\alpha := \alpha(d, c) = 2^{(\log_2 e - c)/(c\,d)} < 1.$$

This implies the following tail estimate.

Lemma 2.29. *For any $\beta > 1$, the probability that SA-forever starts with at least $\lceil \beta \log_{1/\alpha} n \rceil$ controversial rounds is at most*

$$n^{1-\beta}.$$

Proof. The probability for at least this many leading controversial rounds is at most

$$\Pr[C_{\lceil \beta \log_{1/\alpha} n \rceil}] \leq n\alpha^{\lceil \beta \log_{1/\alpha} n \rceil} \leq n\alpha^{\beta \log_{1/\alpha} n} = nn^{-\beta} = n^{1-\beta}.$$

$\square$

We can also bound the expected number of leading controversial rounds in SA-forever, and this bounds the expected number of rounds in SA, because SA terminates upon the first non-controversial round it encounters.

Theorem 2.30. *Let (H, V) be a violator space, $|H| = n$, and $\dim(H, V) = d$. Then the algorithm SA computes a basis of H with an expected number of at most $\mathcal{O}(d \ln n)$ calls to BFA, with at most $\mathcal{O}(d^2)$ constraints each.*

Proof. By definition of C_ℓ, the expected number of leading controversial rounds in SA-forever is

$$\sum_{\ell \geq 1} \Pr[C_\ell].$$

For any $\beta > 1$, we can use (2.28) to bound this by

$$\sum_{\ell=1}^{\lceil \beta \log_{1/\alpha} n \rceil - 1} 1 + n \sum_{\ell = \lceil \beta \log_{1/\alpha} n \rceil}^{\infty} \alpha^\ell = \lceil \beta \log_{1/\alpha} n \rceil - 1 + n \frac{\alpha^{\lceil \beta \log_{1/\alpha} n \rceil}}{1 - \alpha}$$

$$\leq \beta \log_{1/\alpha} n + \frac{n^{1-\beta}}{1 - \alpha}$$

$$= \beta \log_{1/\alpha} n + o(1).$$

This upper bounds the expected number of rounds in SA. In every round of SA one call to BFA is made, using $c\,d^2$ constraints, where $c > \log_2 e$ is constant. □

2.4. Hypercube Partitions

What follows in this section is a small piece about the structure of violator spaces. It concerns the uniqueness of what we call *anti-bases*, and is unrelated to previous sections.

Let H be a finite set. Consider the graph on the vertices 2^H, where two vertices F, G are connected by an edge if they differ in exactly one element, i.e., $G = F \dot\cup \{h\}$, $h \in H$. This graph is a hypercube of dimension $n = |H|$. For the sets $A \subseteq B \subseteq H$, we define $[A, B] := \{C \subseteq H \mid A \subseteq C \subseteq B\}$ and call any such $[A, B]$ an *interval*. A *hypercube partition* is a partition $\mathcal{P}$ of 2^H into (disjoint) intervals.

Let (H, V) be a violator space. We call two sets $F, G \subseteq H$ *equivalent* if $\mathsf{V}(F) = \mathsf{V}(G)$, and let $\mathcal{H}$ be the partition of 2^H into equivalence classes with respect to this relation. We call $\mathcal{H}$ the *violation pattern* of the violator space (H, V).

Before we formulate and prove the *hypercube partition theorem*, we need to introduce some notation. We extend the notion of violator spaces by the concept of anti-basis.

Definition 2.31. *Consider a violator space (H, V). We say that $\bar{B} \subseteq H$ is an* anti-basis *if we have $\mathsf{V}(\bar{B}) \cap F \neq \emptyset$ for all proper supersets $F \supset \bar{B}$. An anti-basis of $G \subseteq H$ is a maximal superset $\bar{B}$ of G with $\mathsf{V}(\bar{B}) = \mathsf{V}(G)$.*

Note that a maximal superset $\bar{B}$ of G such that $\mathsf{V}(\bar{B}) = \mathsf{V}(G)$ is indeed an anti-basis of G. Suppose that there is a set $\bar{B}' \supset \bar{B}$ with $\mathsf{V}(\bar{B}) \cap \bar{B}' = \emptyset$. Locality then decrees that $\mathsf{V}(\bar{B}) = \mathsf{V}(\bar{B}')$, but this contradicts the maximality of $\bar{B}$.

Lemma 2.32. *Consider the violator space (H, V). For any $G \subseteq H$ there is a unique anti-basis $\bar{B}_G$ of G.*

Proof. Suppose that there exist two distinct anti-bases $\bar{B}$ and $\bar{B}'$ of G. Because of $\mathsf{V}(\bar{B}) = \mathsf{V}(\bar{B}')$ and consistency we have that $(\bar{B} \cup \bar{B}') \cap \mathsf{V}(\bar{B}) = (\bar{B} \cup \bar{B}') \cap \mathsf{V}(\bar{B}') = \emptyset$. Therefore, by locality, $\mathsf{V}(\bar{B} \cup \bar{B}') = \mathsf{V}(\bar{B}') = \mathsf{V}(\bar{B})$. Since $\bar{B}$ and $\bar{B}'$ are distinct, it cannot be that $\bar{B} \backslash \bar{B}' = \emptyset$ and $\bar{B}' \backslash \bar{B} = \emptyset$ at the same time. Then, in any case, $|\bar{B} \cup \bar{B}'| > |\bar{B}|$ or $|\bar{B} \cup \bar{B}'| > |\bar{B}'|$ holds, which contradicts the maximality of the anti-bases. $\square$

Corollary 2.33. *Let (H, V) be a violator space, $G \subseteq H$, B_G any basis of G, and $\bar{B}_G$ the unique anti-basis of G. Then for any set F, $B_G \subseteq F \subseteq \bar{B}_G$, F and G are equivalent, i.e., $\mathsf{V}(F) = \mathsf{V}(G)$.*

Proof. This is an immediate consequence of monotonicity (Lemma 2.5).

□

Lemma 2.34. $\mathcal{H}$ *completely determines* (H, V).

Proof. Let $G \subseteq H$. There is a unique anti-basis $\bar{B}_G$ of G, meaning that in $\mathcal{H}$, there is a unique inclusion-maximal superset of G in the same class of the partition. This implies that $\mathsf{V}(G) = \mathsf{V}(\bar{B}_G) = H \setminus \bar{B}_G$, so (H, V) is reconstructible from $\mathcal{H}$.

□

Lemma 2.35. *If* (H, V) *is non-degenerate (unique bases), then* $\mathcal{H}$ *is a hypercube partition.*

Proof. We first show that $\mathsf{V}(B) = \mathsf{V}(B')$ implies $\mathsf{V}(B \cap B') = \mathsf{V}(B \cup B') = \mathsf{V}(B)$. The latter has been shown for the existence of a unique anti-basis. For the former, we argue as follows. Let A be the unique basis of $B \cup B'$. Then $\mathsf{V}(A) = \mathsf{V}(B) = \mathsf{V}(B')$. But then A is also the unique basis of B and B'. It follows that $A \subseteq B \cap B'$, and by locality we get $\mathsf{V}(A) = \mathsf{V}(B \cap B') = \mathsf{V}(B)$.

This argument implies that any partition class $\mathcal{C}$ is contained in the interval $[\bigcap_{C \in \mathcal{C}} C, \bigcup_{C \in \mathcal{C}} C]$. On the other hand, the whole interval is contained in $\mathcal{C}$ by locality, so we are done.

□

Lemma 2.34 and 2.35 together imply that there is an injective mapping from the set of non-degenerate violator spaces to the set of hypercube partitions. It remains to show that the mapping is surjective.

Theorem 2.36. *Any hypercube partition* $\mathcal{P}$ *is the violation pattern of some non-degenerate violator space* (H, V)

Proof. Let $G \subseteq H$, and let $[B, B']$ be the interval containing G. We define $\mathsf{V}(G) = H \setminus B'$ and claim that this is a non-degenerate violator space with violation pattern $\mathcal{P}$. The latter is clear, since $\mathsf{V}(F) = \mathsf{V}(G)$

if and only if $F, G \subseteq [B, B']$. To see the former, we observe that consistency holds because of $G \subseteq B'$. To prove locality, choose $G \subseteq G'$ with $H \setminus B' = \mathsf{V}(G) \cap G' = \emptyset$. In particular, $G' \subseteq B'$, so G' is also in $[B, B']$ and we get $\mathsf{V}(G) = \mathsf{V}(G')$ by definition of V.

It remains to show that the violator space thus defined is non-degenerate. Let B, B' be two sets with $\mathsf{V}(B) = \mathsf{V}(B')$, meaning that they are in the same partition class of $\mathcal{P}$. But then $B \cap B'$ is also in the same class, and we get $\mathsf{V}(B) = \mathsf{V}(B \cap B')$. This implies existence of unique bases. □

2.5. Conclusion

We analyzed Clarkson's algorithm in what we believe to be its most general as well as natural setting. Additionally, we have given the equivalence between non-degenerate violator spaces and hypercube partitions, which could help identifying further applications in computational geometry as well as other fields of computer science. Another major challenge is to develop a sub-exponential algorithm for the third stage, BFA, in the framework of violator spaces (as there already exists for LP's and LP-type problems), or to prove that such an algorithm cannot exist.

The present is a little raft,
floating on the ocean of the past.

3

Simplex Algorithm for Quadratic Programming

This chapter is not going to present original work. It is going to be a summary of the simplex algorithm for quadratic programming that was developed [128] and implemented in CGAL [58] by Gärtner and Schönherr. At later stages Lutz and Wessendorp also contributed to the implementation. The work of Wessendorp is documented in three technical reports. He makes significant improvements, adding support for degeneracies [148], upper bounding [150], and for dealing with a previously unnoticed singularity in the essential linear equation system [149]. Those references are unpublished technical reports that are part of the CGAL documentation. To appreciate the chapters to come, we have to understand how the algorithm works.

For simplicity, let us assume that we are dealing with a non-degenerate quadratic program in standard form (1.2). A problem with inequalities is turned into an equality constrained program by virtue of slack variables. The non-degeneracy assumptions are that (*i*) the rows of A are linearly independent, and (*ii*) the subsystem $A_G x_G = b$ has only solutions for sets $G \subseteq [n]$ with $|G| \geq m$. As mentioned above, the report [148] treats the degenerate case.

3.1. Karush-Kuhn-Tucker Conditions

The foundation for checking whether a feasible solution is optimal are the Karush-Kuhn-Tucker (KKT) conditions, which are derived for general convex optimization problems by requiring that the gradient vanishes at an optimal point [21]. For convex quadratic programs the conditions are necessary *and* sufficient.

Theorem 3.1 (KKT conditions for EQP). *A feasible solution $x^* \in \mathbb{R}^n$ of an equality constrained quadratic program EQP (see (1.2)) is optimal if and only if there exists $\lambda \in \mathbb{R}^m$ and $\mu \in \mathbb{R}^n$, $\mu \geq 0$, such that*

$$c^T + 2x^{*T}D = -\lambda^T A + \mu^T,$$
$$\mu^T x^* = 0.$$

In the algorithm we will consider unconstrained problems of the form (1.3). The following version of the KKT conditions is also known as the method of *Lagrange multipliers*.

Theorem 3.2 (KKT conditions for UQP). *A feasible solution $x^* \in \mathbb{R}^n$ to unconstrained quadratic program UQP (see (3.4)) is optimal if and* *only if there exists* $\lambda \in \mathbb{R}^m$ *such that*

$$c^T + 2x^{*T}D = -\lambda^T A.$$

3.2. Basic Solutions

The linear constraints of (1.2) together with the nonnegativity constraints $x_i \geq 0$ define a polytope $\mathcal{P} = \{p \in \mathbb{R}^n \mid Ap = b, p \geq 0\}$. In linear programming it can be shown that the optimal solution will be found at one of the vertices of $\mathcal{P}$, which are called basic feasible solutions. If a vertex is not an optimal solution, then the constraints at that vertex define a simplicial cone, at least one edge of which leads to a better solution. Hence, the method of pursuing the optimal solution along one of these edges is called *simplex* algorithm. A variable i is called *basic* if $x_i > 0$ in the current solution, and *non-basic* otherwise.

In quadratic programming the optimal solution does not necessarily lie at one of the vertices. Therefore, the definition of a *basis* is more complicated, but it is still characterized by a subset B of the variables that take on a nonzero value in the current solution. All the non-basic variables $N := [n] \backslash B$ will have zero value. Using this assignment for N we can extend any solution to the following problem UQP(B) into a feasible solution of the QP.

Definition 3.3 (QP-Basis). *A subset B of the variables of a quadratic program in standard form (1.2) defines a QP-basis if and only if*

(i) the unconstrained sub-problem

$$\begin{aligned}(UQP(B)) \quad & min \quad c_B^T x_B + x_B^T D_{B,B} x_B \\ & s.t. \quad A_B x_B = b.\end{aligned}$$

(3.4)

has a unique optimal solution $x_B^* > 0$, and

(ii) A_B has full (row) rank, i.e., $\mathrm{rank}(A_B) = m$,

where c_B, $D_{B,B}$, and A_B are the entries of c, D, and A relevant for the variables in B, respectively.

The following theorem gives an upper bound on the basis size. It is crucial to many of the geometric applications described by Schönherr, because it limits the influence of D.

Theorem 3.5 (Theorem 2.6 of [128]). *Every QP-basis B of a quadratic program in standard form (1.2) satisfies*

$$|B| \leq m + \mathrm{rank}(D).$$

Contrary to the traditional way of incorporating the matrix D into the KKT system [152], the previous theorem is the key to reduce the size of the *basis matrix M_B*. This matrix represents the KKT system of the current iteration and consists of a selection of rows and columns from the matrices A and D. With the help of M_B we will be able to perform all the operations necessary during a particular iteration, e.g., checking the optimality conditions or deciding on a search direction. The basis matrix M_B looks as follows,

$$M_B := \left(\begin{array}{c|c} 0 & A_{*,B} \\ \hline A_{*,B}^T & 2D_{B,B} \end{array} \right). \tag{3.6}$$

We are going to see in the next section how to arrive at this formulation.

3.3. Simplex Pivot Step

Here we describe how simplex pivoting works. It is the process of going from one basic solution to another and essentially consists of three steps. The *pricing* is the step in which we check for optimality of the current solution. If that check is negative, it is decided which variable is to enter the basis. This is followed by the *ratio test*, which is to decide which variable has to leave the basis. Finally, in the last step we do the actual update, that is, changing our data structures to reflect the change in basis. Most important, we make the corresponding updates to the basis matrix.

In the following paragraphs we will remain brief and restrict ourselves to the essentials. For a more detailed description we refer to Schönherr [128] and Wessendorp [149].

Pricing

Assume that we are at an iteration with the current basis B, and we would like to check whether adding variable j to the basis can improve the solution. This is done by considering a quadratic program that is restricted to the variables $B \dot\cup \{j\}$. Drawing from the KKT conditions in Theorem 3.1, we derive the following formula for μ_j,

$$c_j + 2{x_B^*}^T D_{B,j} + 2x_j^* D_{j,j} = -\lambda^T A_{*,j} + \mu_j, \qquad (3.7)$$

where $D_{B,j}$ is the jth column of $D_{B,*}$. The vectors λ and x_B^* are obtained by solving the equation system

$$M_B \left(\frac{\lambda}{x_B^*} \right) = \left(\frac{b}{-c_B} \right), \qquad (3.8)$$

if we assume M_B as in equation (3.6).

We solve equation (3.7) for μ_j, and if $\mu_j < 0$, adding the variable j can indeed improve the solution. Only if $\mu_j \geq 0$ for all $j \in [n]\backslash B$ we have arrived at an optimal solution.

To sum up, in the pricing step we have to compute the inverse of M_B, in order to solve equation (3.8), and then evaluate a possibly large number of equations of the form (3.7) involving vector products with the vectors x_B^* and λ.

Ratio Test

At this point we have identified a variable j that is going to enter the basis. Now we would like to find some variable i that will leave the basis while we increase the value of variable j. So we go from basis B, from the previous iteration, to the basis $B \,\dot\cup\, \{j\}\backslash\{i\}$. The non-degeneracy conditions that we postulated at the beginning of this chapter guarantee that the basis matrix is regular for every proper basis according to Definition 3.3, and according to Lemma 2.7 of [128] this is the case. The determination of the leaving variable is done by considering the following unconstrained quadratic program of the form (1.3). Let $\hat{B} := B \,\dot\cup\, \{j\}$.

$$(\mathrm{UQP}_j^t(\hat{B})) \qquad \min \qquad c_{\hat{B}}^T x_{\hat{B}} + x_{\hat{B}}^T D_{\hat{B},\hat{B}} x_{\hat{B}}$$

$$\text{s.t.} \qquad A_{*,\hat{B}} x_{\hat{B}} = b \qquad\qquad (3.9)$$

$$x_j = t,$$

where $t = 0$ is initially zero, and the unique solution $x^*_{\hat{B}}(t)$ to equation (3.9), for each value of t, is determined by the equation

$$M_B \left(\frac{\lambda(t)}{x_B^*(t)} \right) = \left(\frac{b}{-c_B} \right) - t \left(\frac{A_{*,j}}{2D_{B,j}} \right), \qquad (3.10)$$

or equivalently by the equations

$$\begin{pmatrix} \lambda(t) \\ x_B^*(t) \end{pmatrix} = \begin{pmatrix} \lambda \\ x_B^* \end{pmatrix} - t \begin{pmatrix} q_\lambda \\ q_x \end{pmatrix}, \qquad (3.11)$$

$$\begin{pmatrix} q_\lambda \\ q_x \end{pmatrix} := M_B^{-1} \begin{pmatrix} A_{*,j} \\ 2D_{B,j} \end{pmatrix}. \qquad (3.12)$$

The goal is to increase t until some basic variable becomes zero. Without going into further detail, we can see that – similarly to the pricing step – it becomes necessary to compute matrix vector products with the inverse of M_B (see equation (3.12)).

In reality the situation is more complicated for two reasons. First, instead of a basic variable dropping out, it might also happen that the objective function of $(UQP_j^t(\hat{B}))$ will reach a minimum. If this is the case we have to continue with a second step of the ratio test (see [128], page 27-28). Second, it might be that – even if the basis matrix for $B \,\dot{\cup}\, \{j\} \backslash \{i\}$ is regular – both $M_{B \,\dot{\cup}\, \{j\}}$ as well as $M_{B \backslash \{i\}}$ are singular. In this case both update scenarios to get to the new basis matrix by growing and shrinking updates are blocked. This requires a replacement step of variables as described in [149].

Reduced Basis Matrix

We are now going to describe the role of slack variables. Recall that slack variables are introduced to turn inequalities into equalities in the quadratic programming formulation. A crucial ingredient in Schönherr's simplex algorithm is the fact that we can reduce the number of rows and columns considered in the basis matrix. This is described in Section 2.4 of [128]. Let E and S be the sets of indices of equality and inequality constraints respectively. For every constraint in S a slack variable is introduced. Furthermore, let B_O and B_S be the sets of *original* and *slack* basic variables, such that $B = B_O \,\dot\cup\, B_S$. Also, let S_N and S_B be the sets of *non-basic* and *basic* slack variables respectively. These definitions at hand, we can can analyze equation (3.8) again and arrive at the reduced formulation

$$M_B := \left(\begin{array}{c|c} 0 & A_{E\dot\cup S_N, B_O} \\ \hline A^T_{E\dot\cup S_N, B_O} & 2D_{B_O, B_O} \end{array} \right). \tag{3.13}$$

Note that this definition replaces equation (3.6) in the presence of slack variables and reduces the maximal size of the basis matrix to

$$|E| + |S_N| + |B_O| \leq \min\{n, m\} + n,$$

which is a big improvement for the case $m \gg n$. Of course, this condensed formulation has implications for the update procedures, as we will see in Section 3.4.

The Need to Solve Transposed Systems

The last remark that we include in this section concerns the need for solving equation systems determined by the transpose of A. From the previous discussion it should be clear that – in the case of a linear program – the basis matrix looks as follows,

$$M_B = \left(\begin{array}{c|c} 0 & A \\ \hline A^T & 0 \end{array} \right), \tag{3.14}$$

where we left out the selection of particular rows and columns for simplicity. Instead of keeping this matrix as is, the simplex algorithm may only store the matrix A (that is the factorization of A). Therefore, the need arises to solve systems $A\lambda = b$ as well as $A^T x_B^* = -c_B$. Traditionally, in the context of the linear programming simplex algorithm, the former is called FTRAN (*F*orward *TRAN*sformation), while the latter is called BTRAN (*B*ackward *TRAN*sformation). We will adopt the same nomenclature.

Note that these operations are necessary even for a proper quadratic program. The reason for this is that every quadratic program goes through a first phase in order to find an initial feasible solution. This phase I problem is a purely linear optimization problem. At any rate, only A needs to be kept instead of the basis matrix.

3.4. Basis Matrix Updates

After reviewing the course of the simplex algorithm in the last section, let us describe the different types of updates of the basis matrix in this section. In total there are twelve different types of updates. To start, let us list them all (see also Section 6.3.2 of [128], and Section 4 of [149]):

U₁ (QP) An *original* variable *enters* the basis, i.e., B_O is increased by one element.

U₂ (QP) An *original* variable *leaves* the basis, i.e., B_O is decreased by one element.

U₃ (QP) A *slack* variable *enters* the basis, i.e., S_N is decreased by one element.

U₄ (QP) A *slack* variable *leaves* the basis, i.e., S_N is increased by one element.

U₅ (LP) An *original* variable *replaces* an *original* variable in the basis, i.e., one element of B_O is replaced.

U$_6$ (LP) A *slack* variable *replaces* a *slack* variable in the basis, i.e., one element of S_N is replaced.

U$_7$ (LP) An *original* variable *replaces* a *slack* variable in the basis, i.e., B_O and S_N are both increased by one element.

U$_8$ (LP) A *slack* variable *replaces* an *original* variable, i.e., B_O and S_N are both decreased by one element.

U$_{Z_1}$ (QP) An *original* variable *replaces* an *original* variable in the basis, i.e., one element of B_O is replaced.

U$_{Z_2}$ (QP) A *slack* variable *replaces* an *original* variable, i.e., B_O and S_N are both decreased by one element.

U$_{Z_3}$ (QP) An *original* variable *replaces* a *slack* variable in the basis, i.e., B_O and S_N are both increased by one element.

U$_{Z_4}$ (QP) A *slack* variable *replaces* a *slack* variable in the basis, i.e., one element of S_N is replaced.

Note that we differentiate between LP and QP updates. In fact, the first eight updates (U_1-U_8) are already described by Schönherr. The idea was that the replacement type updates (U_5-U_8) are only necessary in the LP case. For the QP case the growing and shrinking updates (U_1-U_4) are sufficient. It was later found by Wessendorp that – also in the QP case – we need a kind of replacement type updates (U_{Z_1}-U_{Z_4}).

Recall that the stored matrix in the LP case is $A_{E \dot{\cup} S_N, B_O}$, while in the QP case we store M_B as in equation (3.13). Figure 3.15 shows the growing and shrinking updates U_1-U_4. The gray elements are the entries to be inserted/deleted respectively. Note that in sub-figure 3.15b

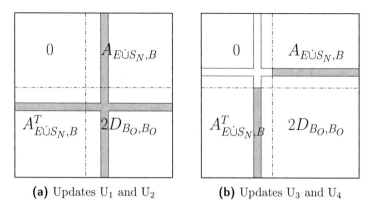

(a) Updates U_1 and U_2 **(b)** Updates U_3 and U_4

Figure 3.15.: The growing and shrinking QP updates (U_1-U_4) are shown.

zero elements are also inserted in the upper left part of M_B. This is indicated by the white elements that are adjoining the gray areas.

Figure 3.16 depicts the replacement updates for the QP case. Note that the replacement areas show a striped pattern in white and gray. This is indicating that those rows and columns are not removed or deleted but replaced. We notice that we do not have to change any entries in the upper left part of sub-figure 3.16b, because they are already zero.

Figure 3.17 compiles updates from QP (3.17a) as well as LP (3.17b). These updates do change the size of the matrix. Updates U_{Z_2} and U_{Z_3} are the most complicated updates in terms of the basis matrix, because they add/delete two rows and two columns each.

Finally, in Figure 3.18 we see the relatively simple LP updates U_5 and U_6. They consist of replacing a row or a column respectively.

In summary, these are all the necessary updates that are used in Schönherr's quadratic programming algorithm. Not all of the updates are equally important, however. Our own experiments indicate

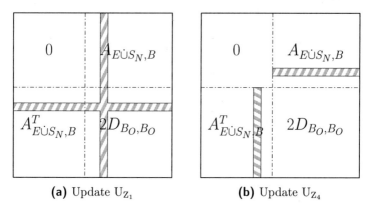

(a) Update U_{Z_1} **(b)** Update U_{Z_4}

Figure 3.16.: QP updates U_{Z_1} and U_{Z_4} are shown. The striped pattern indicates that those elements are replaced.

that – without giving a detailed account – updates involving original variables are a lot more likely than updates involving slack variables. This depends on the particular instance, of course. In an LP problem we do not have any updates of the QP type[9]. As rule of thumb, updates U_1, U_2, U_5, and U_{Z_1} are the ones that are executed the most.

---

[9] The converse is not true, however, because for every QP we need to solve an LP to find an initial feasible solution.

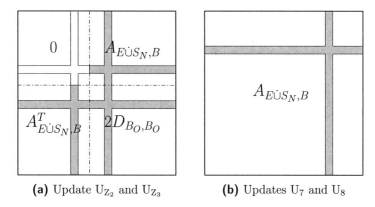

(a) Update U_{Z_2} and U_{Z_3} **(b)** Updates U_7 and U_8

Figure 3.17.: LP update U_7 and U_8.

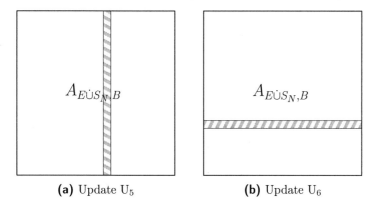

(a) Update U_5 **(b)** Update U_6

Figure 3.18.: LP update U_5 and U_6.

Gentlemen, you can't fight in here!
This is the War Room.

President Merkin Muffley,
from *Dr. Strangelove*
by Stanley Kubrick.

LU factorization

As we have outlined in the previous chapters, the LU factorization is the tool we are going to use to solve linear equation systems that come in two different forms,

$$Ax = b,$$
$$\text{and} \quad A^T x = b. \tag{4.1}$$

The rest of this chapter is structured as follows. First, we are going to give an introduction into the topic of LU factorization in the following section. In Section 4.2 we discuss some classical algorithms and basic properties. This will set the stage for introducing the *integral LU factorization* in Section 4.3. Following up on that, we discuss the issue of sparse matrices in more detail in Section 4.4, and we develop an efficient *update procedure* for the integral LU factorization in Section 4.5.

4.1. Introduction

LU factorization is a well studied topic in linear algebra. It is a key component in several numerical applications, such as solving systems of linear equations (our application), inverting a matrix, or computing the determinant of a matrix. It is closely related to and based on Gaussian elimination. It was first introduced in 1948 by the famous Alan Turing [141]. For a comprehensive treatment see any basic text on linear algebra or matrix computations, such as [119] or [65]. LU factorizations in the context of sparse matrices are considered in an excellent textbook by Davis [37] and another one by Duff, Erisman, and Reid [43].

The tool is in fact so important in practical applications that a great deal of research has been conducted to improve the performance of LU factorizations. One line of work is concerned with conserving the sparsity of matrices, that is, efficiency is boosted by trying to minimize the fill-in of new nonzero elements during the factorization. We are going to have a closer look at this topic in Section 4.4. References are included there.

Another important concern is memory management in systems that have a hierarchical memory layout[10]. This topic is closely related to the question of parallelism in the computation. The matrices to factorize can be huge, so they do not fit into memory as a whole. As soon

---

[10] We omit any mention of literature for vector supercomputers; a topic that has also been studied.

as *paging* starts to occur, memory oblivious algorithms usually grind to a virtual standstill. This is also noticeable in our implementation, and it is definitely an interesting direction for future research. A paper that investigates the *locality of reference* of LU factorizations is, for example, Toledo [140]. Building on the *frontal* approach by Irons [75], whose most important feature is that the computation occurs in only a small part of the matrix (and that only that part of the matrix needs to be kept in memory), the *multifrontal* approach has been developed by Duff and Reid [44]. This method ideally lends itself to parallelization and has been popular in the research community, see for example [35, 36, 38, 39, 92, 76, 5].

A similar approach is to try to reorder the (sparse) matrix such that it consists of blocks that can be independently factored. See for example the paper by Maurer and Wieners [102]. Two recent (and consecutive) PhD theses by Huynh [74] and Maes [94] treat this topic in the context of quadratic programming.

The above considerations of memory locality and parallelism are out of the scope of this thesis, however. Our focus is going to be on the integrality of the computations and a suitable update procedure. Our main goal is to do all the computations over *integral domains*. This is desirable if we want to be able to facilitate efficient and exact computations. If the input to a quadratic programming problem is integer, we want to stay in this realm to avoid doing too many computations over the rationals, which are more expensive. There are other applications for factorizations over integral domains, such as factoring matrices over rings of polynomials, for example. We will introduce the *integral LU factorization* (diLU) and the corresponding routine for solving linear equation systems (sdiLU) in Section 4.3. For more explanations about the number domains considered see Section 4.2.6. As we have found recently, similar results have already been published, but our methods carry some unique traits. For more details see the section about related work concerning the integral LU factorization, Section 4.3.1.

Following up on the integral LU factorization, we develop an efficient update procedure (udiLU) in Section 4.3, that allows us recover the factorization subject to low-rank changes of the original matrix. To the best of our knowledge this is the first result in the realm of integral factorizations. We are building our algorithm on methods that have been developed for the general case. For more details about that, see the related work section concerning the update procedure, Section 4.5.1.

Finally, we are trying to link our results to methods that are used to deal with sparse matrices. This is discussed in Section 4.4. In the case of the integral LU factorization it is possible to incorporate any of the known methods that are used to optimize the factorization procedures. We demonstrate this by applying the *Markowitz rule* to our factorization. However – as we will discuss in Section 4.5.3 – in the case of the update procedure, it is still an open question whether sparse matrices can be accommodated satisfactorily. Sometimes updates do fail. This problem is more pronounced in the case of sparse matrices.

Let us begin with a discussion of some basic properties of and algorithms for the general LU factorization.

4.2. General

In this section we are going to introduce the LU factorization including some basic facts, review different approaches to compute LU factorizations, and close by talking about the number domains that we do our computations in.

4.2.1. Definition

Let $M_n(\mathbb{R})$ be the set of square matrices of order n over the field of the real numbers. Furthermore, let $L_n(\mathbb{R}) \subset M_n(\mathbb{R})$ and $U_n(\mathbb{R}) \subset M_n(\mathbb{R})$ be the set of square *lower triangular* and *upper triangular* matrices respectively. Lower triangular (upper triangular) means, that the matrix has only zero entries above (below) the diagonal. Using this we can formulate the following definition.

Definition 4.2. *Let $A \in M_n(\mathbb{R})$. The matrix A is said to have an* LU *factorization if there exist matrices $L \in L_n(\mathbb{R})$ and $U \in U_n(\mathbb{R})$, such that*

$$A = LU.$$

See Figure 4.3 for an illustration. In general, if such a decomposition exists, it is not unique, but if we prescribe the diagonal elements of, say, L to be equal to 1 then the decomposition is unique (and this is always possible). Furthermore, note that we call a triangular matrix with ones on the diagonal *unit lower triangular* or *unit upper triangular* matrix respectively.

4.2.2. Existence

Let us denote by $A_{i..j,\,k..\ell}$ the sub-matrix that consists of rows i to j of matrix A, but only considering the columns k to ℓ. Writing $A_{i..j}$ we mean the square matrix $A_{i..j,\,i..j}$. If the range of columns or rows represented is the whole range, we may simply write a star, for example, $A_{*,1}$ to denote the first column of A. We could also write

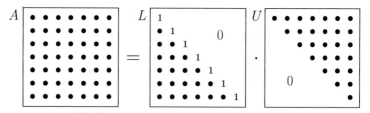

Figure 4.3.: This is an illustration of the matrices L and U of an LU factorization. The dots represent arbitrary elements. Note that L is unit lower triangular.

$A_{*,*}$ to denote the full matrix. And naturally, a single element is denoted by $A_{i,j}$.

The exact conditions for which an LU factorization of A exists, are given in [115]. Note that the following theorem does not only consider square matrices. Even though this is not necessary for our purposes we include it here, because it leads to the conditions for the square case.

Theorem 4.4 (Theorem 1 of [115]). *An arbitrary matrix $A \in \mathbb{R}^{m \times n}$ has an LU factorization if and only if it satisfies the following conditions,*

$$\mathrm{rank}(A_{1..k}) + k \geq \mathrm{rank}(A_{1..k,*}) + \mathrm{rank}(A_{*,1..k}),$$

for all $k \in \{1, \ldots, n\}$.

If the matrix is square and invertible, i.e., it has full rank, then the previous theorem reduces to the following well-known fact.

Corollary 4.5. *An invertible matrix $A \in \mathrm{M}_n(\mathbb{R})$ has an LU factorization if and only if all principal leading sub-matrices of A have full rank.*

Proof. Since A is invertible we have

$$\text{rank}(A_{1..k,*}) = \text{rank}(A_{*,1..k}) = k,$$

for all $k \in \{1, \ldots, n\}$. Then – by Theorem 4.4 – A has an LU factorization if and only if

$$\text{rank}(A_{1..k}) = k,$$

for all $k \in \{1, \ldots, n\}$.   □

4.2.3. Uniqueness

The title of this section is something of a misnomer. In fact, the LU factorization is *not* unique in general. If we require the diagonal elements of the matrix L to be equal to one, as it is usually done in the standard definition, we get the following theorem.

Theorem 4.6 (Theorem 3.2.1 of [65]). *Let $A \in M_n(\mathbb{R})$. If the LU factorization exists, A is non-singular, and the diagonal elements of L are equal to 1, then the LU factorization is unique.*

Proof. Suppose there are two distinct LU factorizations $A = L_1 U_1 = L_2 U_2$. We get that $L_2^{-1} L_1 = U_2 U_1^{-1}$. Note that L_2^{-1} is unit lower triangular and U_1^{-1} is upper triangular. This means that $L_2^{-1} L_1$ is unit lower triangular as well, and $U_2 U_1^{-1}$ is upper triangular. Equality between these two terms is only possible if $L_2^{-1} L_1 = I_n$. Therefore, also $U_2 U_1^{-1} = I_n$, and hence, $L_1 = L_2$ and $U_1 = U_2$, which is a contradiction.   □

As we will see in Section 4.2.5, there is another source of variability. While the conditions of Corollary 4.5 may not be fulfilled for an invertible matrix A, they still can be for some permutation of the rows of A. For all permutations of A that fulfill the conditions we get a different factorization, from which we can recover A. For a simple illustration of this fact, consider the matrix

$$A = \begin{pmatrix} 0 & & \alpha \\ & \cdot^{\cdot^{\cdot}} & \\ \alpha & & 0 \end{pmatrix}.$$

It clearly has an LU factorization if we just invert the order of the rows.

4.2.4. Computation

In this section we will review a few basic methods to transform a matrix into an LU factorization. We will see several methods that are closely related, and try to analyze their suitability for our purposes.

Parts of the following treatment are taken from a summary of matrix algorithms by Timothy Vismor [145]. Furthermore, for more detailed explanations, see the following textbooks [65, 29, 133, 37, 43].

For the rest of this section, let $A = (a_{i,j})_{i,j=1}^{n}$ be the input matrix, consisting of the elements $a_{i,j}$. This is the matrix that we want to factorize,

$$A = \begin{pmatrix} a_{1,1} & \cdots & a_{1,n} \\ \vdots & & \vdots \\ a_{n,1} & \cdots & a_{n,n} \end{pmatrix}.$$

Gaussian Elimination

Arguably the most straightforward way of computing the LU factorization is the Gaussian elimination process. This is also the method upon which the integral LU factorization in Section 4.3 is based. The main idea is to eliminate the sub-diagonal elements of the input matrix A column by column by adding a multiple of a previous row to later rows of the matrix. This procedure is carried out over $n - 1$ iteration steps. The matrix U is completed row by row and L is completed column by column, but in every iteration we update a number of elements in the lower right part of the original A. To see this, consider the following inductive step,

$$\begin{pmatrix} l_{11} & \\ l_{21} & L_{22} \end{pmatrix} \begin{pmatrix} u_{11} & u_{12} \\ & U_{22} \end{pmatrix} = \begin{pmatrix} a_{11} & a_{12} \\ a_{21} & A_{22} \end{pmatrix}, \qquad (4.7)$$

where $l_{11} = 1$ is a scalar. All three matrices are square and partitioned identically. Equation (4.7) implies the following equations,

$$u_{11} = a_{11},$$
$$u_{12} = a_{12},$$
$$l_{21}u_{11} = a_{21},$$
$$l_{21}u_{12} + L_{22}U_{22} = A_{22}.$$

These equations readily yield the values for u_{11}, u_{12}, and l_{21}. Remember that u_{11} is just a scalar. For the lower right part A_{22} we get the following modification,

$$L_{22}U_{22} = A_{22} - l_{21}u_{12}.$$

All elements in the lower right part of the matrix are subjected to a rank-1 update, $A_{22} - l_{21}u_{12}$. We can see that by recursing on that part we ultimately get the full factorization. See Figure 4.8 for an intuitive illustration.

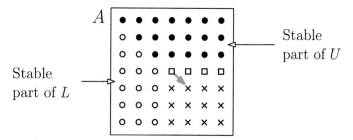

Figure 4.8.: Computational sequence of Gaussian elimination. The crosses indicate elements that are still subject to change. The boxes represent the the pivot row, a multiple of which is going to be subtracted from later rows. The gray arrow represents the progress of the computation. Finally, disks and circles are the final elements of L and U respectively.

If we formulate the update rule of the kth iteration step on an element by element basis, we derive

$$a_{i,j}^{(k+1)} \leftarrow a_{i,j}^{(k)} - \left(\frac{a_{i,k}^{(k)}}{a_{k,k}^{(k)}} \right) a_{k,j}^{(k)}, \tag{4.9}$$

where $i, j > k$. This assumes that we have initialized $U \leftarrow A$. The term $a_{i,k}^{(k)}/a_{k,k}^{(k)}$ describes the effect of eliminating the sub-diagonal elements $a_{i,k}^{(k)}$ for $i > k$. In fact, these multipliers are the elements of the matrix L,

$$l_{i,k} = \frac{a_{i,k}^{(k)}}{a_{k,k}^{(k)}},$$

and we store those elements in the lower left part of $A^{(k+1)}$, because that space is not needed any more. At the end, we can extract the matrix L from there.

Algorithm 3: Gaussian LU factorization

Input: $A \in \mathrm{M}_n(F)$

Output: $L \in \mathrm{L}_n(F)$, $U \in \mathrm{U}_n(F)$, s.t. $LU = A$.

1 **for** $k = 1$ *to* $n - 1$ **do**

2   **for** $i = k + 1$ *to* n **do**

3        $\alpha_{i,k} \leftarrow \frac{a_{i,j}}{a_{k,k}}$;

4        $a_{i,k} \leftarrow \alpha_{i,k}$;

5   **for** $j = k + 1$ *to* n **do**

6           $a_{i,j} \leftarrow a_{i,j} - \alpha_{i,k}\, a_{k,j}$;

7   **end**

8   **end**

9 **end**

10 $L \leftarrow I_n + \mathrm{tril}(A, -1)$;

11 $U \leftarrow \mathrm{triu}(A)$;

Writing everything down we get Algorithm 3, which operates over a field F and does all computations in situ. The instructions in lines 10 and 11 extract L and U from the original space of matrix A. We only have to remember that L has ones on the diagonal. These last operations are optional.

Note the use of the operators $\mathrm{tril}(\cdot)$ and $\mathrm{triu}(\cdot)$. These operators extract the lower or upper triangular part of a matrix respectively. If a second argument is provided it selects the first diagonal that will be extracted, counting the main diagonal as 0, lower diagonals with negative and upper diagonals with positive integer values. As in the algorithm, $\mathrm{tril}(A, -1)$ therefore selects the lower triangular part, leaving out the main diagonal, because -1 indicates that the first sub-diagonal is the first one to be considered.

Other Computation Schemes

Let us mention two popular schemes for computing the LU factorization, which are similar to (yet different from) Gaussian elimination.

If we assume that L is unit lower triangular and U is upper triangular, we can simply apply the definition of matrix multiplication to obtain

$$a_{i,j} = \sum_{p=1}^{\min\{i,j\}} l_{i,p} u_{p,j}, \tag{4.10}$$

where $1 \leq \{i,j\} \leq n$. If we rearrange the terms in this equation we get

$$l_{i,j} = \frac{1}{u_{j,j}} \left(a_{i,j} - \sum_{p=1}^{j-1} l_{i,p} u_{p,j} \right), \tag{4.11}$$

for $i > j$, and

$$u_{i,j} = a_{i,j} - \sum_{p=1}^{i-1} l_{i,p} u_{p,j}, \tag{4.12}$$

for $j \geq i$. Using these two equations in the correct order, we can compute all elements in closed from, that is, without doing multiple updates on the unstable elements of the matrix. We include the implementation of this algorithm that is known as *Doolittle's algorithm* in Appendix A.4. Briefly speaking, the algorithm computes one row after another. For a quick intuition of the computation order, please refer to Figure 4.15a. Note that Doolittle's algorithm can handle matrices that are not invertible. The resulting matrix L will always be a matrix of full rank with ones on the diagonal, whatever the rank of A may be. If the rank of A is not full, then this fact is only exhibited by the matrix U that will have the same rank as A. In particular U will have $n - \text{rank}(A)$ trailing zero rows.

Similarly, if we assume that U is a unit upper triangular and L is

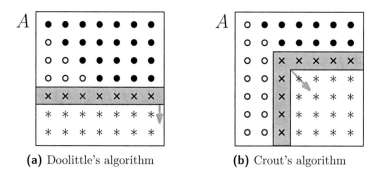

(a) Doolittle's algorithm   **(b)** Crout's algorithm

Figure 4.15.: Computational sequences of Crout's and Doolittle's algorithms. As in Figure 4.8 the circles and disks represent stable
parts of L and U respectively. The crosses are the elements
that are computed in the current iteration. The stars are
original elements of A. They are completely untouched, up
to this stage.

a lower triangular matrix, we get a similar procedure, based on the
formulas

$$l_{i,j} = a_{i,j} - \sum_{p=1}^{i-1} l_{i,p} u_{p,j}, \tag{4.13}$$

for $i \geq j$, and

$$u_{i,j} = \frac{1}{l_{j,j}} \left(a_{i,j} - \sum_{p=1}^{i-1} l_{i,p} u_{p,j} \right), \tag{4.14}$$

for $i < j$. This results in an algorithm that is known as *Crout's
algorithm*. Again, for the implementation see Appendix A.5 and for
the computational sequence see Figure 4.15b. In a nutshell, Crout's
algorithm computes one column of L and one row of U after another.

These two algorithms, Crout's and Doolittle's algorithms, have an
important advantage. They require less intermediate precision to get
more accurate results using floating point arithmetic. To substantiate
this seemingly obscure claim, consider how the elements are computed

in Crout's and Doolittle's algorithm. In (4.11) and (4.12) we can see, for example, that the major share of the work is done by computing an inner product between two vectors. If we keep adding products of corresponding elements of the two vectors to a variable of twice the precision of the input numbers, we can easily obtain the numerically accurate result. Rounding that is applied at the end of this process is unlikely to result in a large numerical error. This way of accruing numerical errors is sometimes called *inner product accumulation*, and generally leads to better numerical stability [133]. By contrast, looking at the Gaussian elimination process, we realize that the final value of an element is modified over the course of several iterations. Even if we compute the intermediate results exactly, we must always round the result at the end of each update if we want to keep it in the same precision that is used for the input. This process that we might call by the name of *partial sum accumulation* therefore suffers from potentially bad numerical behavior.

There are disadvantages to Crout's and Doolittle's algorithms too, however. First, it is more difficult to implement row and column interchanges (LU pivoting, see Section 4.2.5), which is an essential tool for stability and efficiency in the case of sparse inputs [69]. Second, and this is the more severe restriction for us – as we will see later – the elements of the result that we are computing share common divisors across rows. This means that all the elements of some row i are expressed as rational numbers with the same common divisor. Elements from different rows do not have the same common divisor. Implementing the formulas of Crout and Doolittle, we therefore have to compute common divisors. This is either inefficient, or leads to increased growth of the numbers, or both. We want to avoid this. Furthermore, since we will compute our results with arbitrary precision over the fraction field of an integral domain, we will also be able to avoid the problem of accruing errors from which partial sum accumulation usually suffers.

Eta File

In the context of the simplex method, the *eta file* – more expressively known as the *product form of the inverse* [34] – used to be a popular tool for keeping the matrix L as a sequence of extremely sparse elimination matrices. The advantage of this method is that a matrix in this form is trivial to invert and – at least conceptually – lends itself well to updates of the matrix that has to be factored. If that happens the eta file simply grows, accounting for the changed elements. If the eta file gets too large, one can attempt a re-factorization. It has been proved that the LU factorization of a sparse matrix is usually sparser than the product form. For a more detailed discussion see also [137]. Nevertheless, we are briefly going to review whether this method is suitable for our purposes. For that we need to define a few tools.

If $x = (x_1, \ldots, x_n)^T$ is an arbitrary vector, we define the $n \times n$ matrix $L_k(x)$ to be the matrix that has ones on the diagonal, and fulfills the equation

$$L_k(x)\, x = (x_1, \ldots, x_k, 0, \ldots, 0)^T,$$

where $1 \leq k \leq n$. In other words, multiplying x with $L_k(x)$ leaves the first k entries of x unchanged and renders the remaining entries equal to 0. It is easy to see that

$$L_k(x) = I_n - l_k(x)\, e_k^T,$$

where e_k is the kth unit vector, and $l_k(x) = \frac{1}{x_k}(0, \ldots, 0, x_{k+1}, \ldots, x_n)^T$.

We call a matrix of this form *Gaussian elimination matrix*. Of course, this definition is only sound if $x_k \neq 0$. One can check that the inverse of $L_k(x)$ is

$$(L_k(x))^{-1} = I_n + l_k(x)\, e_k^T,$$

which is easy to compute, of course. Note that a Gaussian elimination matrix as well as its inverse are lower triangular. For example, $L_k(x)$ is of the form

$$L_k(x) = \frac{1}{x_k} \begin{pmatrix} x_k & & & & \\ & \ddots & & & \\ & & x_k & & \\ & & x_{k+1} & & \\ & & \vdots & \ddots & \\ & & x_n & & x_k \end{pmatrix}, \qquad (4.16)$$

where all missing entries are equal to 0. Only the kth column is filled below the diagonal.

We can now express Gaussian elimination in terms of these elimination matrices. We successively multiply A with Gaussian elimination matrices. Each of these multiplications removes the nonzero elements below the diagonal in one column of A. So, the first step is computing

$$A^{(2)} := L_1(A_{*,1})\, A,$$

where the Gaussian elimination matrix depends on the first column of A. Let us call this intermediate result $A^{(2)}$, which is the input to the second elimination round (we silently assumed $A^{(1)} := A$). Note that $A^{(2)}$ has only zero entries below the diagonal in the first column. For this first step to go through we need $a_{11} \neq 0$.

The second step does the following,

$$A^{(3)} := L_2(A^{(2)}{}_{*,2})\, A^{(2)},$$

where $A^{(2)}{}_{*,2}$ is the second column of the matrix $A^{(2)}$. For this to work, we need $a_{22}^{(2)} \neq 0$. If we do this for all the first $n-1$ columns of A, we finally arrive at an upper triangular matrix $U := A^{(n)}$. Since there is no ambiguity, let us write $L_k := L_k(A^{(k)}{}_{*,k})$. Then the elimination procedure can be conveniently written as

$$L_{n-1} \cdot \ldots \cdot L_1\, A = U,$$

under the requirement that $a_{ii}^{(i)} \neq 0$ for $1 \leq i < n$. Finally, if we define $\hat{L} := L_{n-1} \cdot \ldots \cdot L_1$ and $L := \hat{L}^{-1}$, we get the desired equality $A = LU$.

Unfortunately, this popular method in linear programming is not suited for our purposes, because it seems unlikely that the growth of numbers can be bounded in the context of the increasing number of elimination matrices.

4.2.5. Pivoting

In practice Theorem 4.4 and Corollary 4.5 are of little interest, because they assume that we are not allowed to permute the rows and columns of the matrix. The following theorem is much more useful.

Theorem 4.17. *An invertible matrix $A \in \mathrm{M}_n(\mathbb{R})$ allows for a LUP-decomposition, i.e.,*

$$PA = LU,$$

where P is a permutation matrix.

Proof. Applying Gaussian elimination with row interchanges achieves the desired result.

Using the same definitions as above, we may find ourselves in elimination step i, in which $a_{i,i}^{(i)} = 0$. The problem is, of course, that the Gaussian elimination matrix for this column is not well defined.

Since A is invertible, there must be some element $a_{j,i}^{(i)} \neq 0$ for $i < j \leq n$. We have to exchange row i with row j in the matrix $A^{(i)}$. Assume that this is achieved by multiplying (from the left) with the permutation matrix P_i. Note that the sub-matrix $A^{(i)}{}_{i..n,\,1..i-1}$ is the zero matrix, so applying P_i does not disturb the work we have done so far.

At the end, the elimination procedure leaves us with

$$L_{n-1} P_{n-1} \cdot \ldots \cdot L_1 P_1 \; A = U,$$

where the P_i are permutation matrices and the L_i are Gaussian elimination matrices. One can invest five minutes of thought to see that we can move the P_i to the right, commuting with the L_i. To be more precise, $PL = L'P$, where L' has the elements of its active column permuted according to P, though the diagonal remains unchanged. Note that L' is still lower triangular. Therefore, we have

$$L'_{n-1} \cdot \ldots \cdot L'_1 \; P_{n-1} \cdot \ldots \cdot P_1 \; A = U,$$

and we can set $L := (L'_{n-1} \cdot \ldots \cdot L'_1)^{-1}$ and $P := P_{n-1} \cdot \ldots \cdot P_1$ in order to get the statement of the theorem. The matrix $L'_{n-1} \cdot \ldots \cdot L'_1$ is lower triangular and invertible, because the individual L'_i are. Therefore, also L is lower triangular. Also, $P_{n-1} \cdot \ldots \cdot P_1$ is a permutation matrix, because the product of permutation matrices is always a permutation matrix. $\qquad\square$

Note that P is an orthogonal matrix, and therefore $P^{-1} = P^T$, and thus $A = P^T LU$. Whether we view P to be a permutation that acts on A or on LU is really just a matter of definition. Note that multiplying A by a permutation matrix from the left corresponds to row permutations. Therefore, what this type of decomposition procedure does is, go through the ith column of the matrix $A^{(i)}$ and select one of the yet unchosen rows to find the pivot element for the next elimination step. This is called *partial pivoting*. Note that this type of pivoting is even sufficient to factorize singular matrices if we take a little more care about the computation order. The result will be a regular lower triangular matrix L and a singular upper triangular matrix U (including zero rows at the bottom).

Full Pivoting

Partial LU pivoting can be enhanced by considering the columns that have not yet been handled as well. This means that, in the kth step of the elimination procedure, there are $(n-k+1)^2$ potential elements to choose the pivot element from. This is an enhancement in the sense that it allows for more sophisticated pivoting strategies, which do not generate as much fill-in for sparse matrices. The fill-in minimizing techniques described in Sections 4.4.3 and 4.4.4 will make use of full pivoting. Also, it can help to improve numerical stability. It is not necessary, however, for the existence of a LUP factorization according to Theorem 4.17.

Solving an Equation System with LU

Suppose that we want to solve for the vector x in the equation system $Ax = b$, where x and b are vectors of length n. Using the LUP-decomposition of A we can divide this task into two steps. The equation now reads $P^T LUx = b$ and can be split up,

$$U x = y, \qquad (4.18)$$

$$P^T L y = b. \qquad (4.19)$$

Both equations can be solved efficiently because of the diagonal structures of L and U. The operation of solving an equation system as above is called FTRAN (Forward TRANsformation). This term has developed historically and is used in the literature.

Similarly, we will also have to consider equation systems of the following type, $x^T A = b^T$. This is equivalent to $x^T P^T LU = b^T$, and therefore

$$x^T P^T L = y^T, \qquad (4.20)$$

$$y^T U = b^T. \qquad (4.21)$$

These equations give us an efficient way to solve this problem. This operation is called BTRAN (Backward TRANsformation).

4.2.6. Number Types

Numerical errors are a real concern in practical implementations of LU factorization. In the mathematical description of a factorization routine we do not have to care about errors, because computations are assumed to be exact (usually over the real numbers). In real-world implementations on actual computer hardware, however, the most natural way – and fast for that matter – of doing computations is to use one of the built-in finite precision number types such as `float` or `double`. In other words, one uses floating-points arithmetic that adheres to a standard like the IEEE standard 754 [2]. The main problem with that course of action is that we *will* make mistakes in our computations due to the finite precision of those number types.

In our treatment and implementation of the LU factorization we take a different approach. Since one of the main features of the quadratic programming solver of CGAL is its ability to be instantiated with a custom number type, and to do computations *exactly*, we will investigate this aspect further. In simple terms, our goal is to provide a factorization procedure that does not need divisions (at least not in the general sense).

Even though most of the implementation details will be discussed in Chapter 5, the following paragraphs will anticipate certain aspects. We hope that this ordering of topics delivers clarity and motivation to our exposition, to outweigh the deficits in consistency. Let us therefore embark on a little excursion about algebraic number types, group theory and their practical counterparts in CGAL.

Algebraic Number Types

In abstract algebra, we can distinguish the following (incomplete) chain of class inclusions:

$$
\begin{aligned}
\text{Commutative rings} &\supset \text{integral domains} \\
&\supset \text{unique factorization domains} \\
&\supset \text{Euclidean domains} \\
&\supset \text{fields}
\end{aligned}
\tag{4.22}
$$

We assume that the definitions of these terms are known to the reader. They can be looked up in any standard book on abstract algebra. Here, we will be satisfied by stating the key properties and differences between these structures.

Briefly speaking, a *commutative ring* is a set R equipped with two binary operations $+$ and $\cdot$ that combine any two elements from the ring. These operations are called addition and multiplication. The structure $(R, +)$ is required to be an Abelian group, and $(R, \cdot)$ a commutative monoid (more descriptively, it is a commutative group without an inverse). If we additionally require that there are no non-trivial zero divisors, i.e., no two elements $a, b \in R$, $a \neq 0$ and $b \neq 0$ such that $a \cdot b = 0$ (where 0 is the identity element of addition) we get an *integral domain*. This is the structure that we base our LU factorization on. That is, we require division $a \div b$ only when it is known that there exists an element $q \in R$ such that $b \cdot q = a$. The integer numbers $\mathbb{Z}$ are the prototypical example of an integral domain. The division as described before corresponds to integral division.

In fact, the integer numbers also belong to the classes of *unique factorization domains* and *Euclidean domains*. The former distinguishes itself by the existence of a unique factorization within the structure, e.g., prime factorization. The latter assumes the existence of a well-defined division with remainder.

The integer numbers are not a field though. What sets *fields* apart from all of the above is that they do allow for a general division operation. The structure is closed under division. Among the most common examples for fields are the set of real numbers $\mathbb{R}$ and the set of rational numbers $\mathbb{Q}$.

Let us round up the algebraic description by saying that all the inclusions of equation (4.22) are proper, even though, for a practical realization of some number type, the only distinction we will have to make is between fields and Euclidean domains, of which rational and integer numbers are examples of, respectively.

CGAL Number Types

In a practical implementation of some algorithm, it is essential which operations are actually implemented for a certain number type. So, it still makes sense to talk about such a fine grained distinction as the above. In CGAL these structures are present as concepts. Figure 4.23 shows the dependencies between the concepts. For example, the concept `IntegralDomain` requires the implementation of a function `integral_division`. Since the two concepts `Field` and `UniqueFactorizationDomain` refine the concept of `IntegralDomain` they also need to postulate that function. Basically, that is what the arrows in the figure indicate. Realizations of these concepts are called models, i.e., we say that some number type `t` is a model of some concept `c` if it implements all the postulated data members and functions of `c`. If `c` refines some parent concept `p`, naturally, `t` is also a model of `p` by transitivity.

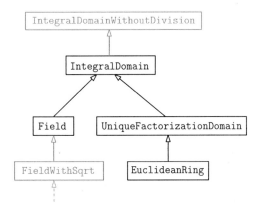

Figure 4.23.: Algebraic structure concepts implemented in CGAL.

We can notice a difference between the algebraic description and the concepts in CGAL. In particular, `Field` is not a refinement of

EuclideanRing but of IntegralDomain directly. This is not to say that the algebraic interpretation of equation (4.22) is faulty. The difference in CGAL is simply a practical design decision. Some of the properties and tools of unique factorization domains and Euclidean rings are not worth implementing for fields, because they can be replaced by division to provide more efficient algorithms. In the words of the CGAL manual, algebraic foundations package[11]:

> ... this is because most ring-theoretic notions like greatest common divisors become trivial for Fields. Hence we see Field as a refinement of IntegralDomain and not as a refinement of one of the more advanced ring concepts. If an algorithm wants to rely on greatest common divisor or remainder computation, it is trying to do things it should not do with a Field in the first place.

We did not talk about the gray items in the figure yet, because they are not of major importance for us but shown for completeness. IntegralDomainWithoutDivision postulates an integral domain in the algebraic sense, but it has no implementation of integral_division. Also, there are several refinements of Field, the first of which is FieldWithSqrt that, naturally, postulates an operation for taking square roots. Others are omitted here.

These concepts need actual realizations, of course. Let us therefore mention the most important number types that we will be working with. Note that we restrict ourselves to the programming language C++. We have already mentioned the finite-precision types float and double, which are standardized in IEEE 754 [2].

---

[11] Section 4.2 of http://www.cgal.org/Manual/latest/doc_html/cgal_manual/Algebraic_foundations/Chapter_main.html

Other than that, CGAL offers its own number types and wrapper for several well-known external arithmetic libraries. Most of the experimental results in Section 6 were conducted with the wrapper classes for the GNU Multiple Precision Arithmetic Library (GMP) types. The most important wrappers for us are the types `Gmpzf` and `Gmpq`. The former is an arbitrary-precision floating point type that is based on the type `mpz`, which is an GMP arbitrary size integer type, to represent the mantissa and the exponent. Of course, `Gmpz`, as the name suggests, is also based on `mpz` but only needs one integer to store its contents. Both `Gmpfz` and `Gmpz` are models of `EuclideanRing`. The type `Gmpq` is a model of `Field`. Instances are kept as a rational number, where nominator and denominator have arbitrary size.

Other Number Types

There is one more GMP type that is supported by CGAL. There is `Gmpfr`, which is a fixed-precision floating point type. This is no better that `float` or `double` in some sense, because – if we want to allow for exact computations – we need to assure that the precision is sufficient that no rounding will take place. Given Theorem 4.50, which asserts item (iii) of Claim 4.24 in particular, we are in a position to give a sensible upper bound on the size of the numbers. But the main purpose of a type like `Gmpfr` would be defeated. We would have to make all instances of `Gmpfr` large enough to accommodate the largest numbers occurring during the computation. This would ensure exact computations, but it would also mean that many computations take place using *too much* precision. For determining the necessary precision as tightly as possible we have to do preprocessing steps, i.e., compute the determinant of the matrix to be factored. This is not only of the same order of complexity than computing the desired factorization; it amounts to essentially doing the same thing.

There is also a set of exact number types provided by CGAL itself. Among those are `Quotient`, which can be used to represent rational numbers given any number type that can represent integers. Also, similar to `Gmpzf`, there is the internal type `MP_float`. These internal types are generally less efficient than the GMP types. They are there to provide the possibility of exact computations independent on any external libraries.

Last but not least, other external number type libraries supported by CGAL are LEDA[12] and CORE[13]. Let us, however, not get into any detail about these, because we did not do any testing using these types.

Summary

In this section we took a glimpse ahead into some of the underlying implementation details of the factorization algorithm we are going to present in the next section. We have outlined the tools that are necessary to do exact computations, and we have explained how those number types correspond to their algebraic counterparts. In this spirit, we hope to have set the stage, and explained the motivations for an integral LU factorization, which does all computations over an integral domain.

---

[12] http://www.algorithmic-solutions.com/as_html/products/leda/products_leda.html

[13] http://www.cs.nyu.edu/exact/core/

4.3. Integral LU factorization

In this section we describe a way to efficiently factorize a matrix over an integral domain into a form which enables integral computations as well as sparse optimization. Let us start by listing the properties that we want this decomposition to have.

Claim 4.24. *A matrix $A \in \mathrm{M}_n(\mathcal{I})$ over some integral domain $\mathcal{I}$ can be factored into matrices L, U_i, L_i, and U, such that the following properties hold:*

(i) $A = LU_i^{-1} = L_i^{-1}U$, *where L and L_i are lower triangular, and U and U_i are upper triangular.*

(ii) *During the computation only integral division is used. That is, the only division operation we need is dividing two numbers α and β, already knowing that the result will be integer. In the language of group theory, we might say that there has to be an integral element γ such that $\gamma \cdot \beta = \alpha$.*

(iii) *The entries of L, U_i, L_i, and U are given as rational numbers, where the denominators are never larger than $\det(A)$. The encoding size of any number computed during the factorization is bounded by $\mathcal{O}(\langle\det(A)\rangle)$.*

(iv) *The decomposition pair L_i and U caters to the FTRAN operation (equations (4.18) and (4.19)) in such a way that the entries of the final solution vector are given as rational numbers with a common denominator of $\det(A)$.*

(v) *The decomposition pair (L, U_i) caters to the BTRAN operation (equations (4.20) and (4.21)) in the same way as the pair (L_i, U) caters to FTRAN (see previous item).*

The notation $\langle\cdot\rangle$ designates the encoding size of a number; it is described in Appendix A.1.

When doing Gaussian elimination on a matrix, it is not difficult to modify the standard algorithm so as not to use divisions. This usually leads to rapid growth of the entries though, as discussed in [67, 142]. One of the main concerns of the algorithm that we are going to propose is to prevent this growth in the entries.

4.3.1. Related Work

We have developed our work based on an exposition by Grötschel, Lovász, and Schrijver, in which they argue that the Gauss-Jordan elimination process[14] of a rational matrix can be executed in strongly polynomial time (see [67], Section 1.4). This was originally proved by Edmonds [47].

In summary, our procedure differs from the one described in [67] insofar as the latter does not give an explicit formulation for the matrix L. It implicitly computes the same matrix U but then carries on with a summary argument that $A^{-1}b$ can be computed by following through with the Gauss-Jordan elimination. There is no provision to extract L from this procedure.

There is a series of papers, however, that essentially derives the same result as we do. Therein, the properties that we describe in Claim 4.24 are subsumed by the term *fraction-free* LU factorization. We have become aware of this work only after developing our own results.

There are slight differences in our objectives and algorithms. Let us give an overview of those publications and point out some issues.

---

[14] The Gauss-Jordan elimination process is an extension of regular Gaussian elimination, where we also eliminate the elements above the diagonal in A. This basically amounts to computing $A^{-1}b$, except for the fact that we still have to divide by the diagonal elements.

The main motivation for that parallel line of work comes from having to solve linear algebra systems over some integral domain, arising from computations with polynomials or differential equations. The paper [110] gives a good overview of some applications. Two of the first publications in this direction were written by Bareiss [11, 12], proposing greatest common divisor reduction on the elements of the matrix. While this prevents the growth of elements, it leads to an increased number of operations. Later the topic was picked up again in several papers [143, 88, 31, 159, 108] in the context of applications in *threat detection* and *robot control.*

Most recently, the subject is treated in [46] and [159, 158]. The former discusses the case when the matrix to factorize is singular. The latter is similar to our own. Essentially, the same algorithm is derived, but the elements on the diagonal are chosen differently. Also, the algorithm proposed by Zhou and Jeffrey [158] does compute L and not L^{-1}, which requires a little more work when doing the forward substitution. In particular, an integral $LD^{-1}U$ factorization is derived. While it is also possible to solve for A *and* for A^T with this setup, the numbers that are computed during the solution process (see equations (4.18)-(4.21)) get larger than in our procedure. On the downside, in our approach, we have to spend a little more time to compute the factorization of A^T along the way[15].

4.3.2. Algorithm (diLU)

In this section we are going to present our own factorization algorithm. Basically, the algorithm is straight-forward Gauss elimination

---

[15] Note that the effort for computing the factorization of A *and* A^T is considerably less than twice what we have to spend to obtain either individually; see Section 4.3.5.

Algorithm 4: Double-Integral-LU (diLU)

Input: $A \in M_n(\mathcal{I})$
Output: $(L_\imath, U), (L, U_\imath) \in L_n(\mathcal{I}) \times U_n(\mathcal{I})$,
         s.t. $L_\imath^{-1} U = L U_\imath^{-1} = A$.

1   $P \leftarrow [A, I_n]$;
2   $q_1 \leftarrow 1$;
3   **for** $k = 1$ **to** $n - 1$ **do**
4      $q_{k+1} \leftarrow P_{[k,k]}$;
5   **for** $i = k + 1$ **to** n **do**
6   **for** $j = k + 1$ **to** $n + k$ **do**
7             $P_{[i,j]} \leftarrow (q_{k+1} P_{[i,j]} - P_{[i,k]} P_{[k,j]}) \div q_k$;
8   **end**
9   **for** $j = 1$ **to** k **do**
10         $P_{[j,n+i]} \leftarrow (q_{k+1} P_{[j,n+i]} - P_{[k,i]} P_{[j,n+k]}) \div q_k$;
11   **end**
12         $P_{[i,n+i]} \leftarrow q_{k+1}$;
13   **end**
14 **end**
15 $L_\imath \leftarrow \mathrm{tril}\big(P_{[*, n+1..2n]}\big)$;
16 $U \leftarrow \mathrm{triu}\big(P_{[*, 1..n]}\big)$;
17 $L \leftarrow \mathrm{tril}\big(P_{[*, 1..n]}\big)$;
18 $U_\imath \leftarrow \mathrm{triu}\big(P_{[*, n+1..2n]}\big)$;

---

with some modifications. One could say that we apply it twice, once to the matrix A and once more to the matrix A^T. Conceptually, the two decomposition pairs $(L_\imath, U)$ and $(L, U_\imath)$ are derived from decomposing A and A^T independently, but as we will show one can make considerable savings by deriving the two factorizations at the same time.

To keep things simple, let us assume that all the principal leading sub-matrices of A have full rank. In other words, A is invertible and according to Corollary 4.5 it allows for an LU factorization without pivoting. We will come back to this topic of pivot rules in Section 4.4.

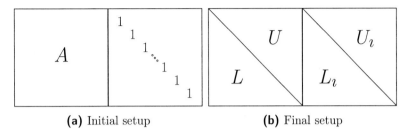

(a) Initial setup   (b) Final setup

Figure 4.25.: An illustration of the initial and the final setup of the working matrix in the diLU algorithm.

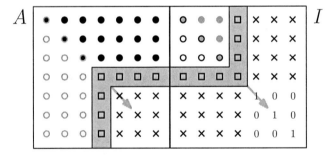

Figure 4.26.: Computational sequence of the diLU algorithm. The crosses indicate elements that subject to change in the current iteration. The boxes are the pivot elements. The disks and circles are final elements of L_i, U_i, L, and U. The decomposition pair (L_i, U) is represented in black, and the decomposition pair (L, U_i) is represented in gray. Figure 4.25 shows the initial and final setup.

The same assertions also apply to the matrix A^T, because it has the same principal leading sub-matrices only transposed, which does not affect the rank, of course. Having established this precondition, we can easily formulate the algorithm without burdening it by row and possibly column permutations.

Recall that $\div$ denotes integral division. Let us have a closer look at how the algorithm works. It follows the Gaussian elimination steps closely. First however, we initialize a $n \times 2n$ matrix P as a working matrix. The left part of P is the matrix A and the right part is the identity matrix of order n. In the end P will contain the factorizations (L_i, U) and (L, U_i); see Figure 4.25. We save space by overwriting the contents of A, but in order to store both factorizations we need the additional space on the right-hand side, of course. For an illustration of the computational sequence, see Figure 4.26.

For simplicity, we continue describing the governing rules of the factorization of A into (L_i, U) only, that is, we disregard the factorization of A^T for the time being. It should be clear that the algorithm is symmetric in the sense that we obtain the same decomposition pairs if we start out with A^T (with the roles of lower and upper triangularity reversed), and that all the following arguments apply equally. We will come back to the complexity – and in particular the savings – of computing both factorization pairs at the same time in Section 4.3.5.

In any iteration, at the *beginning* of the outer loop (line 3 to 14), we have the following invariant,

$$A = (L_i^{(k)})^{-1} U^{(k)}, \tag{4.27}$$

where $L_i^{(k)}$ is the right part of P, and $U^{(k)}$ is the left part of P. These two matrices have the following form,

$$
L_i^{(k)} := \left(\begin{array}{c|c} L_{11}^{(k)} & 0 \\ \hline L_{21}^{(k)} & L_{22}^{(k)} \end{array} \right) = \left(\begin{array}{ccccc|ccc} q_1 & & & & & & & \\ * & q_2 & & & & & 0 & \\ \vdots & \ddots & \ddots & & & & & \\ * & \cdots & * & q_{k-1} & & & & \\ \hline * & \cdots & & * & q_k & & & \\ \vdots & & & \vdots & & \ddots & & \\ * & \cdots & & * & 0 & & q_k & \end{array} \right),
$$

and

$$
U^{(k)} := \left(\begin{array}{c|c} U_{11}^{(k)} & U_{12}^{(k)} \\ \hline 0 & U_{22}^{(k)} \end{array} \right) = \left(\begin{array}{cccc|ccc} q_2 & * & \cdots & * & * & \cdots & * \\ & q_3 & \ddots & \vdots & \vdots & & \vdots \\ & & \ddots & * & & & \\ & & & q_k & * & \cdots & * \\ \hline & & & & * & \cdots & * \\ & & 0 & & \vdots & & \vdots \\ & & & & * & \cdots & * \end{array} \right),
$$

where $*$ stands for an unnamed entry. That is, an entry designated by $*$ could still be zero, but it is not inherently zero, like the off-diagonal entries in the identity matrix, for example. Note that, for $k = 1$ we have $U_{22}^{(1)} = A$ and $L_{22}^{(1)} = I_n$.

For $k > 1$, the parts $L_{11}^{(k)}$, $U_{11}^{(k)}$, and $U_{12}^{(k)}$ are already stable. In particular, $L_{11}^{(k)}$ is lower triangular, and $U_{11}^{(k)}$ is upper triangular. The contents of $L_{21}^{(k)}$, $L_{22}^{(k)}$, and $U_{22}^{(k)}$ are still to be manipulated in the remaining iterations. Ultimately, $L_{11}^{(n)}$ and $U_{11}^{(n)}$ contain the full factorization L_t and U. To understand what happens in one iteration of the outer loop, consider $U_{22}^{(k)}$,

$$U_{22}^{(k)} = \begin{pmatrix} u_{1,1}^{(k)} & \cdots & u_{1,n-k+1}^{(k)} \\ \vdots & & \vdots \\ u_{n-k+1,1}^{(k)} & \cdots & u_{n-k+1,n-k+1}^{(k)} \end{pmatrix}.$$

The next step will be to eliminate the nonzero entries in the first column below $u_{1,1}^{(k)}$. In regular Gauss elimination we subtract a multiple of the first row from all the other rows, according to

$$u_{i,j}^{(k+1)} \leftarrow u_{i,j}^{(k)} - \frac{u_{i,1}^{(k)}}{u_{1,1}^{(k)}} \, u_{1,j}^{(k)}, \qquad (4.28)$$

for all $i, j = 2, \ldots, n - k + 1$. Let us assume for a moment that this is exactly what we are doing. Consider the equation system

$$U^{(k)}x = L_i^{(k)}b, \qquad (4.29)$$

for some arbitrary vector b and vector of unknowns x. According to invariant (4.27), this system has the same solution as $Ax = b$. If we add multiples of one equation to another equation, we do not change the solution of the system. This is what the Gaussian update rule (4.28) is based on. An important consequence is that we can get rid of the unwanted division operation. Multiplying by $u_{1,1}$ gives the following rule,

$$u_{i,j}^{(k+1)} \leftarrow u_{1,1}^{(k)} \, u_{i,j}^{(k)} - u_{i,1}^{(k)} \, u_{1,j}^{(k)}. \qquad (4.30)$$

Note that this operation keeps invariant (4.27) intact for the next iteration, because we can relay the changes in the equation system to the right hand side $L_i^{(k)}b$ by implementing the same kind of update to the matrices

$$L_{21}^{(k)} = \begin{pmatrix} l_{1,1}^{(k)} & \cdots & l_{1,k-1}^{(k)} \\ \vdots & & \vdots \\ l_{n-k+1,1}^{(k)} & \cdots & l_{n-k+1,k-1}^{(k)} \end{pmatrix},$$

and

$$L_{22}^{(k)} = \begin{pmatrix} q_k & & 0 \\ & \ddots & \\ 0 & & q_k \end{pmatrix}.$$

The update for $L_{21}^{(k)}$ is analogue to equation (4.30),

$$l_{i,j}^{(k+1)} \leftarrow u_{1,1}^{(k)} l_{i,j}^{(k)} - u_{i,1}^{(k)} l_{1,j}^{(k)}, \tag{4.31}$$

for all $i = 2, \ldots, n - k + 1$ and $j = 1, \ldots, k - 1$. For $L_{22}^{(k)}$ we get the update rules

$$\left(L_{22}^{(k+1)}\right)_{i,1} \leftarrow -u_{i,1}^{(k)} q_k,$$
$$\left(L_{22}^{(k+1)}\right)_{i,i} \leftarrow u_{1,1}^{(k)} q_k,$$

for all $i = 2, \ldots, n - k + 1$. In principle, this is the same as equation (4.31), but it simplifies due to the special structure of $L_{22}^{(k)}$.

The only thing left out yet are the integral division operations that are applied in lines 7, 10, and 12 of diLU (Algorithm 4). We will defer this discussion to Section 4.3.3. For now, let us assume that all of those divisions are integral. Then we get the final update formulas

$$u_{i,j}^{(k+1)} \leftarrow \frac{u_{1,1}^{(k)} \, u_{i,j}^{(k)} - u_{i,1}^{(k)} \, u_{1,j}^{(k)}}{q_k}, \qquad (4.32)$$

$$l_{i,j}^{(k+1)} \leftarrow \frac{u_{1,1}^{(k)} \, l_{i,j}^{(k)} - u_{i,1}^{(k)} \, l_{1,j}^{(k)}}{q_k}, \qquad (4.33)$$

for all $i, j = 2, \ldots, n - k + 1$ (the latter also holds for $j = 1$), and

$$\left(L_{22}^{(k+1)}\right)_{i,\,1} \leftarrow -u_{i,1}^{(k)}, \qquad (4.34)$$

$$\left(L_{22}^{(k+1)}\right)_{i,\,i} \leftarrow u_{1,1}^{(k)}, \qquad (4.35)$$

for all $i = 2, \ldots, n - k + 1$. In the case of (4.34) and (4.35) it is even obvious that there is an integral division involved. Finally, we realize that $q_{k+1} = u_{1,1}^{(k)}$. The updates (4.32), (4.33), and (4.34) are done in lines 6 to 8. The update (4.35) is done in line 12.

Note that the lines 9 to 11 are part of computing the second factorization pair (L, U_ι) for A^T. Part of the computations done in lines 6 to 8 pertain to that as well. We are not going to go into further detail why Algorithm 4 computes the second factorization pair as well. Let us just say that the preceding arguments also apply in the transposed context of A^T.

Lemma 4.36. *The factorizations computed by Algorithm 4 are correct, i.e., $A = LU_\iota^{-1} = L_\iota^{-1}U$, where L and L_ι are lower triangular, and U and U_ι are upper triangular. This lemma corresponds to property* (i) *of Claim 4.24.*

Proof. Except the fact that we have not proved yet that all the divisions used during the reduction process are integral, it should be clear from the previous description that the proposed algorithm is a scaled form of the Gaussian elimination process, and that it does therefore compute a correct factorization. If the integral domain is actually a field, we have completely proved the lemma already, because we can just carry out the divisions in field arithmetic. That is the case for the rational numbers, for example. The other case is covered by Lemma 4.37. □

4.3.3. Integrality

In Claim 4.24 we have listed a number of properties that we wish to confer on the diLU factorization. In the sections to come those properties are going to be derived. Property (i) basically just says that the decomposition is correct and that we can compute it with the algorithm given. We have argued that point in Section 4.3.2 already, Lemma 4.36, except for the fact that the divisions applied should be integral. Therefore, our next target is property (ii).

Property (ii) of Claim 4.24 states that the only divisions executed are integral division. This concerns lines 7 and 10 of Algorithm 4 where we divide by q_k.

Lemma 4.37. *For any invertible matrix $A \in \mathrm{M}_n(\mathcal{I})$, all divisions executed by Algorithm 4 are integral (property (ii) of Claim 4.24 is fulfilled). Furthermore, the encoding size of any number computed during the factorization is bounded by $\mathcal{O}(\langle \det(A) \rangle)$.*

Proof. Recall the matrices U_{22} and L_{21} from iteration k,

$$
U^{(k)} = \left(\begin{array}{c|c} U_{11}^{(k)} & U_{12}^{(k)} \\ \hline 0 & U_{22}^{(k)} \end{array} \right),
$$

$$L_i^{(k)} = \left(\begin{array}{c|c} L_{11}^{(k)} & 0 \\ \hline L_{21}^{(k)} & L_{22}^{(k)} \end{array} \right),$$

$$U_{22}^{(k)} = \left(\begin{array}{ccc} u_{1,1}^{(k)} & \cdots & u_{1,n-k+1}^{(k)} \\ \vdots & & \vdots \\ u_{n-k+1,1}^{(k)} & \cdots & u_{n-k+1,n-k+1}^{(k)} \end{array} \right),$$

$$L_{21}^{(k)} = \left(\begin{array}{ccc} l_{1,1}^{(k)} & \cdots & l_{1,n-k+1}^{(k)} \\ \vdots & & \vdots \\ l_{n-k+1,1}^{(k)} & \cdots & l_{n-k+1,n-k+1}^{(k)} \end{array} \right).$$

The main idea is to express the elements of $U_{22}^{(k)}$ as follows,

$$u_{i,j}^{(k)} = \frac{p_{i,j}^{(k)}}{s^{(k)}}, \tag{4.38}$$

where $s^{(k)} = \det(U_{11}^{(k)})$. This should hold inductively for all values $k = 1, \ldots, n$. Naturally, we initialize with $p_{i,j}^{(1)} := a_{i,j}$ and $q^{(1)} := 1$. The claim is that both $p_{i,j}^{(k)}$ and $s^{(k)}$ are elements from the integral domain, i.e., we do not need to extend our computations to the field of fractions of the integral domain. In fact, we claim that the values encountered are minors of the original matrix A. Let us use the following index sets $K := \{1, \ldots, k-1\}$, $I := \{1, \ldots, k-1, k+i-1\}$, and $J := \{1, \ldots, k-1, k+j-1\}$. It is easy to see that $U^{(k)}{}_{K,K} = U_{11}^{(k)}$, as per our definition. Also, note that $U^{(k)}{}_{I,J}$ is an upper triangular matrix with the entry $u_{i,j}^{(k)}$ in the lower right-hand corner and $U_{11}^{(k)}$ in the upper left part. Therefore, we can develop the determinant of $U^{(k)}{}_{I,J}$ as

$$\det\left(U^{(k)}{}_{I,J}\right) = u_{i,j}^{(k)} \, \det\left(U^{(k)}{}_{K,K}\right).$$

Rearranging we get

$$u_{i,j}^{(k)} = \frac{\det\left(U^{(k)}{}_{I,J}\right)}{\det\left(U^{(k)}{}_{K,K}\right)}. \tag{4.39}$$

Now, assume that we are doing the standard Gaussian elimination update as described by update rule (4.28), and the analogue extension of that rule to the right-hand side part of our working matrix,

$$l_{i,j}^{(k+1)} \leftarrow l_{i,j}^{(k)} - \frac{u_{i,1}^{(k)}}{u_{1,1}^{(k)}}\, l_{1,j}^{(k)}.$$

In other words, the updates we are doing amount to adding multiples of earlier rows to later rows. If we recall that the determinant of a matrix does not change if we add multiples of one row to another row, we conclude that

$$\det\left(U^{(k)}{}_{I,J}\right) = \det\left(A_{I,J}\right),$$
$$\det\left(U^{(k)}{}_{K,K}\right) = \det\left(A_{K,K}\right).$$

That simplifies our target formula (4.39) for the representation of the elements to

$$u_{i,j}^{(k)} = \frac{\det\left(A_{I,J}\right)}{\det\left(A_{K,K}\right)}. \tag{4.40}$$

Note that this already tells us – if we manage to express the elements of our factorization as in equation (4.40) – the encoding size of numbers computed will be bounded by $\mathcal{O}(\langle\det(A)\rangle)$. This takes care of the second statement of the lemma.

To finish the proof about the integrality of the division, consider plugging equation (4.38) into the update rule (4.28). We derive the following formulas for $p_{i,j}^{(k+1)}$ and $s^{(k+1)}$,

$$p_{i,j}^{(k+1)} = \frac{p_{i,j}^{(k)}\, p_{1,1}^{(k)} - p_{i,1}^{(k)}\, p_{1,j}^{(k)}}{s^{(k)}}, \tag{4.41}$$

$$s^{(k+1)} = p_{1,1}^{(k)}. \tag{4.42}$$

Note that $p_{1,1}^{(k)} = \det(A_{\{1,\dots,k+1\}, \{1,\dots,k+1\}}) = s^{(k+1)}$. This is the reason why we know that (4.41) holds. Because we know that $p_{i,j}^{(k+1)}$ is itself an integer number, we may conclude that the division is an integral division.

This concludes the proof. Let us point out a few more things though. First, we have omitted the last few steps for deriving the formulas for L_i. It should be clear, however, that the same arguments apply. Second, note that we have basically derived the update rule (4.32) again. The values $s^{(k)}$ are the same as the values that we have designated by q_k before. We intentionally named them differently, because we wanted to make the point that that was not clear a priori. Last, note that in our algorithm we are only explicitly storing the values $p_{i,j}^{(k+1)}$ of representation (4.38). The values q_k are implicitly stored on the diagonal (of U as well as L_i).   □

4.3.4. Solving Linear Systems (sdiLU)

To make our treatment of the integral LU factorization complete we have to describe the final step – and ultimate goal – of the elimination process. That is, we have to describe how to solve a linear equation system using the factorization that we have obtained in previous sections. What we are going to show in this section corresponds to the items (iv) and (v) of Claim 4.24. For simplicity, we will only discuss item (iv) in detail. The case of the transposed matrix in (v) is completely analogous with the roles of L and U exchanged. For convenience, let us gather all the relevant equations that are necessary to find the solution to $Ax = b$. To start with let us assume full LU pivoting, so there is a row permutation matrix P and a column permutation matrix C,

$$P\,A\,C = L_i^{-1}\,U, \tag{4.43}$$

$$P^T L_i^{-1}\,y = b, \tag{4.44}$$

$$U\,C^T\,x = y. \tag{4.45}$$

Equation (4.43) describes the output we get from our factorization algorithm diLU (Algorithm 4), that is the factorization pair L_i and U, and also the permutation matrices P and C. Note that the L_i^{-1} exists over the field of fractions of our input domain $Frac(\mathcal{I})$, but we will not really have to compute it. This is because equation (4.44) transforms to $y = L_i\,P\,b$, where y is an intermediate vector that we have to compute. This equation is analogous to equation (4.19) but tuned to our particular factorization. Finally, equation (4.45) is analogous to equation (4.18) but with column permutation added. This is the second step we have to take. By doing a backward substitution on U we can obtain the solution vector x. Note that the two permutation matrices can be easily absorbed by the intermediate vector y and the solution vector x. So, for clarity we will drop them, or in other words assume that $P = C = I_n$. After transformations and dropping the permutations, the two equations (4.44) and (4.45) simplify to

$$y = L_i\,b, \tag{4.46}$$

$$x = U^{-1}\,y. \tag{4.47}$$

These two equation are what we compute in the following algorithm.

Algorithm 5: Solve-Double-Integral-LU (sdiLU)

Input: $L_i \in L_n(\mathcal{I})$, $U \in U_n(\mathcal{I})$ (obtained from diLU(A)), $b \in \mathcal{I}^n$

Output: $x \in \mathcal{I}^n$, $d \in \mathcal{I}$, s.t. $Ax/d = b$ on $Frac(\mathcal{I})$

1   $y \leftarrow L_i b$;   ▷ Handle equation (4.46)

2   $q \leftarrow \mathrm{diag}(U)$;   ▷ Extract q_i's

3   $d \leftarrow q_n$;   ▷ $d = \det(A)$

4   $x_n \leftarrow y_n$;   ▷ Last entry of solution

5   **for** $i = n - 1$ **to** 1 **by** -1 **do**   ▷ Handle equation (4.47)

6      $t \leftarrow y_i d$;

7   **for** $j = i + 1$ **to** n **do**

8         $t \leftarrow t - x_j U_{i,j}$;

9   **end**

10     $x_i \leftarrow t \div q_i$;   ▷ Divide by q_i's

11 **end**

Equation (4.46) is handled in line 1, while the backward substitution of equation (4.47) is dealt with in the for loop starting at line 5, and the one instruction just before the loop (line 4). Note that this latter instruction could seamlessly be integrated into the loop itself. Due to the fact that the inner loop is empty for $i = n$, however, we would bluntly multiply with d and then divide by d again. We can save us that trouble. In the remaining initialization, lines 2 and 3, we extract the common denominators from the diagonal of U in line 2, the last one of which deserves a special variable because it is part of the output. Recall that the last diagonal entry of U is the determinant of A, and it will be the the common denominator for the elements of the solution vector x. We call this value d and extract it in line 3.

As the final output we get the vector $x \in \mathcal{I}^n$ and $d \in \mathcal{I}$, such that $\tilde{x} := \frac{x}{d} \in Frac(\mathcal{I})$ is the unique solution to $A\tilde{x} = b$ (recall that we assume A to be invertible in $Frac(\mathcal{I})$). Of course, all of this is standard lore in linear algebra if we were to use arithmetic operations over a field. In particular, if the division in line 10 of Algorithm 5 were a field operation. Here, however, the division is integral, and it is the main goal of the remainder of this section to prove this. Consider the following lemma.

Lemma 4.48. *In Algorithm 10, assuming input from an integral domain $\mathcal{I}$, and preprocessing by the diLU procedure (Algorithm 4), all arithmetic operations employed are over $\mathcal{I}$, in particular all division operations are integral.*

Proof. As for all the additions, subtractions, and multiplications, it is clear that the integral domain provides these operations. We only have to prove that t is always a multiple of q_j in line 10 of the algorithm. To see this consider Cramer's rule. It states that we can express the solution to the system $A\tilde{x} = b$ as ratios of determinants. For the values of $\tilde{x}$ it follows that

$$\tilde{x}_i = \frac{\det(A_i)}{\det(A)}, \tag{4.49}$$

for $i = 1, \ldots, n$, where A_i is the matrix formed by replacing the ith column of A by the vector b. From Lemma 4.36, Lemma 4.37, and the above discussion we have already convinced ourselves that the result we compute is correct over $Frac(\mathcal{I})$, and that so far we have only operated over $\mathcal{I}$. Now, we start to compute the values for $\tilde{x}$ one by one using a representation that assumes $d := \det(A)$ as the common denominator for the entries. Looking at the formula for one particular value,

$$\tilde{x}_i = \frac{y_i - \sum_{j=i+1}^n \frac{x_j}{d} U_{i,j}}{q_i}$$

$$= \frac{d\,y_i - \sum_{j=i+1}^n x_j U_{i,j}}{d\,q_i}$$

we realize that

$$x_i := \frac{d\,y_i - \sum_{j=i+1}^n x_j U_{i,j}}{q_i}$$

must be an integral value, and that this is in fact exactly the computation that we are doing in the algorithm. Here, we assume that all values of x_j for $j > i$ are already known. This concludes the proof. □

4.3.5. Complexity of diLU and sdiLU

Computing the factorization (diLU)

Computing the complexity of Algorithm 4 is straightforward. Let us count the number of multiplications, the number of subtractions, and the number of integral divisions. We do not care about the initializing statements or the final assignment operations, because they can be implemented efficiently without any effort. The whole algorithm can be executed *in situ*, overwriting the input matrix. Of course, the caller needs to provide twice the space of the input matrix, but if he does that, the initializing and finalizing statements are purely conceptual.

Let us get back to the number of multiplications N_* then. Summation over the nested loops yields

$$N_* = \sum_{k=2}^{n}\sum_{i=k}^{n}\left(\sum_{j=k}^{n+k-1} 2 + \sum_{j=1}^{k-1} 2\right)$$

$$= 2\sum_{k=2}^{n}\sum_{i=k}^{n}(n+k-1)$$

$$= 2\left(\frac{2}{3}n^3 - \frac{1}{2}n^2 - \frac{1}{6}n\right).$$

The number of divisions $N_{\div}$ and the number of subtractions N_- is the same, because each of the two innermost loops contains exactly one of these operations. Therefore we get

$$N_{\div} = N_- = \sum_{k=2}^{n}\sum_{i=k}^{n}\left(\sum_{j=k}^{n+k-1} 1 + \sum_{j=1}^{k-1} 1\right)$$

$$= \frac{2}{3}n^3 - \frac{1}{2}n^2 - \frac{1}{6}n.$$

Computing the total number of operations we get

$$N = N_* + N_- + N_{\div} = \frac{8}{3}n^3 - 2n^2 - \frac{2}{3}n,$$

for doing the factorization of both A and A^T in the fashion we described.

This has to be contrasted with the expense it takes to independently compute both factorization pairs (L_t, U) and (L, U_t). In fact, the respective counts are obtained by omitting the second loop on the innermost level in Algorithm 4. That gives $2n^3 + \mathcal{O}(n^2)$ for the total number of operations for one factorization pair. Hence, the savings we are making by using the proposed procedure are $\frac{4}{3}n^3 + \mathcal{O}(n^2)$, or roughly one third of the total number of operations.

In 1974 it was shown by Bunch and Hopcroft [23] that LU factorization (as well as matrix inversion) can be achieved substantially faster if one employs fast matrix multiplication algorithms. To be concrete, if we are given an algorithm that can multiply two matrices of size $n \times n$ in time $\mathcal{O}(n^{\omega})$ for some $\omega < 3$, then we can theoretically get the same result for LU factorization. At the time of writing this thesis, the fastest known matrix multiplication algorithm seems to be the yet unpublished algorithm by Virginia Williams [151], which constitutes an incremental improvement upon the famous Coppersmith-Winograd algorithm [30]. The reported running time depends on $\omega \approx 2.3727$. The constants hidden in the Landau notation are so large, however, that this type of algorithm does not lend itself to a practical implementation. The asymptotic advantage will only play out for matrices that are too large to be handled by modern day computers. Therefore, this algorithm can be of no further interest to us at this point.

There is a relatively simple yet clever algorithm by Volker Strassen that is successfully used in practice, however. It was the first algorithm to achieve a running time of $o(n^3)$ for matrix multiplication. In his original paper [135] Strassen derives a running time of $4.7n^{\log_2 7}$ arithmetic operations. In the terminology of the previous paragraph that means $\omega = \log_2 7 \approx 2.8$. Compared to the roughly $2n^3$ operations that straightforward matrix multiplication needs, we conclude that Strassen's algorithm starts paying off when $n > 2^{\frac{\log_2 4.7 - 1}{3 - \log_2 7}} \approx 84$.

On the one hand, this threshold is already quite large for our application where the real computational effort comes from iterating through a large number of bases of moderate size, and not necessarily from solving huge systems. On the other hand, our algorithm does not really rely on multiplicative schemes such as the application of eta files[16]. Also, it has been reported that factorization schemes that rely on Strassen's algorithm suffer from loss of numerical stability [10].

In conclusion, incorporating the method of Strassen is not likely to give a big performance boost, given the size of the matrices involved, and even more important, it is not clear in any way how the integral methods could be adapted to work with such a scheme.

Solving the equation system (sdiLU)

In the same way as in the previous paragraph, we can easily obtain the number of multiplications, divisions, and subtractions by analyzing Algorithm 5,

$$N_* = \sum_{j=1}^{n-1} \left(1 + \sum_{k=j+1}^{n} 1 \right)$$
$$= \frac{1}{2}n^2 + \frac{1}{2}n - 1.$$

The number of subtractions is $N_- = \frac{1}{2}n^2 - \frac{1}{2}n$ and the number of divisions is $N_\div = n - 1$. The total count of operations is therefore

$$N = N_* + N_- + N_\div = n^2 + n - 2.$$

This is less than what we need if we have to multiply the right-hand side by an explicit inverse, which gives $2n^2 - n$ multiplications and additions. There are no divisions involved in the latter, however.

---

[16] Recall that Section 4.2.4 introduces eta files.

4.3.6. Summary

Theorem 4.50. *The double integral LU factorization, as computed by diLU (Algorithm 4) for some matrix $A \in M_n(\mathcal{I})$ over an integral domain $\mathcal{I}$, $\det(A) \neq 0$, fulfills all the properties listed in Claim 4.24.*

Proof. Item (i) (correctness) was implicitly proved by deriving the factorization from a scaled version of the Gaussian elimination scheme in Section 4.3.2. Item (ii) (integrality of the factorization) and (iii) (bound on the encoding size of numbers) have been established in Lemma 4.37. Item (iv) (solving linear systems) is shown to be possible using only integral divisions in Lemma 4.48.

Finally, item (v) (decomposition pair for A^T) is not explicitly proved, but is treated implicitly in all the expositions we mentioned above.   □

4.4. Sparse Matrices

So far we have disregarded one of the major design goals of the integral LU factorization, in favor of the numerical and algorithmic considerations. We will make up for that in the present section. As we have already stated, one of the main reasons for replacing the inverse of the basis matrix with an LU factorization is to allow for better treatment of sparse inputs. While the inverse of a sparse matrix is *not* sparse in general, there are much better techniques available to keep the LU factorization sparse. For an excellent in-depth treatment of the topics mentioned in this chapter, see Chapters 6 and 7 of the textbook by Davis [37].

Note that there is no formal definition of a sparse matrix. It simply designates a matrix that has *mostly* zero entries.

4.4.1. Fill-In

One of the main characteristic of the quality of a sparse factorization method is the amount of *fill-in* it generates.

Assume that the matrix A is uniquely factorized by partial pivoting into $PA = LU$, where P is a permutation matrix. Recall that we can enforce uniqueness of the factorization by requiring that all the diagonal elements of L be 1.

Definition 4.51. *Let $PA = LU$ be the unique LU factorization of A, where L is unit lower triangular. The fill-in is defined as the set of nonzero elements of $L + U$ that are zero in PA.*

Let us give an example. Consider the matrix A,

$$
A = \begin{pmatrix} 2 & 2 & 1 & 1 \\ 1 & 2 & 0 & 0 \\ 1 & 0 & 2 & 0 \\ 1 & 0 & 0 & 4 \end{pmatrix}, \tag{4.52}
$$

and the following factorization,

$$
PA = \begin{pmatrix} 1 & 0 & 0 & 0 \\ 0 & 1 & 0 & 0 \\ 0 & 0 & 1 & 0 \\ 0 & 0 & 0 & 1 \end{pmatrix} \begin{pmatrix} 2 & 2 & 1 & 1 \\ 1 & 2 & 0 & 0 \\ 1 & 0 & 2 & 0 \\ 1 & 0 & 0 & 4 \end{pmatrix}
$$

$$
= \begin{pmatrix} 1 & 0 & 0 & 0 \\ \frac{1}{2} & 1 & 0 & 0 \\ \frac{1}{2} & -1 & 1 & 0 \\ \frac{1}{2} & -1 & -1 & 1 \end{pmatrix} \begin{pmatrix} 2 & 2 & 1 & 1 \\ 0 & 1 & -\frac{1}{2} & -\frac{1}{2} \\ 0 & 0 & 1 & -1 \\ 0 & 0 & 0 & 2 \end{pmatrix} = LU. \tag{4.53}
$$

We can see that P is the identity matrix. No pivoting was necessary to obtain the given factorization. We can also see that $L + U$ is a fully dense matrix,

$$L + U = \begin{pmatrix} 3 & 2 & 1 & 1 \\ \frac{1}{2} & 2 & \boxed{-\frac{1}{2}} & \boxed{-\frac{1}{2}} \\ \frac{1}{2} & \boxed{-1} & 2 & \boxed{-1} \\ \frac{1}{2} & \boxed{-1} & \boxed{-1} & 3 \end{pmatrix}.$$

If we compare the circled elements with the respective positions in A, we notice that these are exactly the *new* nonzeros, and therefore constitute the fill-in of factorization (4.53). With the help of matrix A we can show that the elimination order matters when it comes to fill-in. Using a different elimination order, we obtain the following factorization,

$$P_2 A = \begin{pmatrix} 0 & 0 & 0 & 1 \\ 0 & 1 & 0 & 0 \\ 0 & 0 & 1 & 0 \\ 1 & 0 & 0 & 0 \end{pmatrix} \begin{pmatrix} 2 & 2 & 1 & 1 \\ 1 & 2 & 0 & 0 \\ 1 & 0 & 2 & 0 \\ 1 & 0 & 0 & 4 \end{pmatrix}$$

$$= \begin{pmatrix} 1 & 0 & 0 & 0 \\ 1 & 1 & 0 & 0 \\ 1 & 0 & 1 & 0 \\ 2 & 1 & \frac{1}{2} & 1 \end{pmatrix} \begin{pmatrix} 1 & 0 & 0 & 4 \\ 0 & 2 & 0 & -4 \\ 0 & 0 & 2 & -4 \\ 0 & 0 & 0 & -1 \end{pmatrix} = L_2 U_2.$$

We have exchanged the last row with the first row in the elimination order. Conveniently, that reduced the amount of fill-in, as we can see from $L_2 + U_2$,

$$
L_2 + U_2 = \begin{pmatrix}
2 & 0 & 0 & 4 \\
1 & 3 & 0 & \boxed{-4} \\
1 & 0 & 3 & \boxed{-4} \\
2 & 1 & \frac{1}{2} & 0
\end{pmatrix}.
$$

Only two elements are still encircled. Note that we have to compare the entries with $P_2 A$ to reflect the changed pivot order. Also note that there is one new zero entry on the diagonal. Let us disregard this, however, for two reasons. First, we are not interested in the diagonal elements at any rate. If the factorization exists (which we may assume), we will have nonzero elements on the diagonal of both L_2 and U_2, because those are exactly the pivot positions. In the current example, we get an accidental cancellation in $L_2 + U_2$ because $(L_2)_{4,4} = -(U_2)_{4,4}$, but this is not of interest when considering the fill-in. Second, we have to attest that sometimes cancellations may happen *during* the elimination process. So, elements that were originally nonzero may become zero. This phenomenon is relatively rare, however, and compared to the occurrence of fill-in it is negligible.

Now, to draw on our example one last time, let us demonstrate that *perfect elimination* is possible – if we allow not only for the rows but also the columns to be permuted. Consider inverting both the order of the rows and the columns in matrix A. This is easily achieved by applying the permutation matrix

$$P_3 = \begin{pmatrix} 0 & 0 & 0 & 1 \\ 0 & 0 & 1 & 0 \\ 0 & 1 & 0 & 0 \\ 1 & 0 & 0 & 0 \end{pmatrix}$$

from both sides. That is, we obtain the factorization

$$P_3 A P_3^T = \begin{pmatrix} 4 & 0 & 0 & 1 \\ 0 & 2 & 0 & 1 \\ 0 & 0 & 2 & 1 \\ 1 & 1 & 2 & 2 \end{pmatrix}$$

$$= \begin{pmatrix} 1 & 0 & 0 & 0 \\ 0 & 1 & 0 & 0 \\ 0 & 0 & 1 & 0 \\ \frac{1}{4} & \frac{1}{2} & 1 & 1 \end{pmatrix} \begin{pmatrix} 4 & 0 & 0 & 1 \\ 0 & 2 & 0 & 1 \\ 0 & 0 & 2 & 1 \\ 0 & 0 & 0 & \frac{1}{4} \end{pmatrix} = L_3 U_3.$$

The remarkable realization is that *no fill-in* is generated at all by this ordering. Indeed, matrices with a sparsity pattern such as we find in A are sometimes referred to as *arrow matrices*. They are the prototypical example of how elimination order matters when it comes to fill-in. If a matrix (with a fixed ordering) does not produce any fill-in, we call it a *perfect elimination matrix*.

4.4.2. Minimizing Fill-In

As we have seen in Section 4.4.1, we can considerably reduce the size of our factorization by permuting the original matrix in such a way that fill-in is minimized. By size we mean the number of nonzero elements that we have to store. In the example we have seen, the right permutation makes a big difference. If we generalize the example of the arrow matrix in equation (4.52) to size $n \times n$, we end up with the difference between having to store n^2 or only $3n - 2$ entries. So a natural question is this: Can we always find the pivot ordering that minimizes the fill-in efficiently? The answer to that question is negative (unless $P=NP$), because the problem is NP-complete. In this section we are briefly going to review what is known about hardness, before we go on to techniques that handle the problem well in practice. Note that there is a comprehensive chapter about fill-reducing orderings in [37].

For the symmetric realm – that is for the Cholesky factorization – it was proved in 1981 by Yannakakis [154] that computing the minimum fill-in is NP-complete. The proof is a reduction from the *optimal linear arrangement problem*, that was shown to be NP-complete by Garey and Johnson [55]. The reduction takes a few intermediate graph theoretic steps that are interesting in their own right. Most important, the minimum fill-in problem is shown to be equivalent to determining the minimum number of edges that have to be added to make a given graph chordal. Note that perfect elimination matrices as well as chordal graphs can be recognized in linear time [126].

Since this problem is difficult to solve in general, we may revert to heuristics that are successful in practice. Let us briefly describe some of those in the following two sections.

4.4.3. Markowitz Pivot Rule

One of the earliest elimination reducing heuristics was proposed by Markowitz [98]. It is particularly noteworthy that Markowitz received the Nobel prize[17] for economics for his seminal contributions to portfolio theory in 1990 (see also [96]), part of which is his fill-in reducing work on LU factorizations.

The rule is easily formulated as follows. Consider the first Gaussian elimination step during the factorization of the matrix A. The rule states that we should choose the element $a_{i,j}$ that minimizes the Markowitz count

$$(r_i - 1)(c_j - 1), \tag{4.54}$$

where r_i is the number of nonzero entries in row i, and c_j is the number of nonzero entries in column j. Ties may be resolved arbitrarily. The matrix is permuted such that the element $a_{i,j}$ is moved to the uppermost leftmost position.

Even though this rule is not optimal in all instances, it proves to be very successful and efficient in practice. It is implemented as the LU pivoting strategy in our implementation as well.

Duff, Erisman, and Reid give a more detailed investigation of this pivoting rule [43]. Most of those considerations are of minor interest to us, however, because they deal with increasing numerical stability, which we have less trouble maintaining using exact arithmetic.

---

[17] http://www.nobelprize.org/nobel_prizes/economics/laureates/1990/markowitz-autobio.html

4.4.4. Other Heuristics

A large body of research has been devoted to find more elaborate pivoting strategies than the one described in the previous section. Most of them are based on the *elimination tree* of a matrix A, which is derived from the graph representation G_A. The graph G_A is simply defined as $V(G_A) := [n]$, and $E(G_A) := \{(i,j) \mid a_{i,j} \neq 0\}$. The elimination tree depends on the elimination order, and is not easily defined. Roughly speaking, we remove all edges from G_A that do not lead to fill-in. See Davis [37] for a comprehensive treatment.

Two of the most popular methods are *minimum degree* and *nested dissection*. The former is a greedy strategy, trying to unravel the elimination tree from its leaves (always picking a vertex of minimum degree). The Markowitz rule can also be seen as such a degree minimizing method. For more advanced algorithms, see [139, 3, 4, 41, 40], for example. The latter method called nested dissection tries to cleverly cut up the elimination tree starting from its root; see [61, 62, 42, 120]. These methods are particularly successful in factorizing matrices that arise from 2D or 3D finite element formulations. For more explanations and references see Chapter 7 of [37].

It might be of interest to investigate some of these techniques in the context of our own factorization methods. For the present implementation we were satisfied with the results of the Markowitz rule. We did see a marked improvement stepping from partial pivoting (which only guarantees the existence of a factorization) to Markowitz pivoting. These experiments are undocumented.

4.5. Update of Integral LU factorization

One of the major concerns when solving linear equation systems in the context of linear and quadratic programming as well as in other applications is a fast update procedure. Assume that we have already computed the solution to the system

$$Ax = b,$$

where $A \in M_n(\mathcal{I})$. The question is whether and how we can exploit the work that has been done already to solve a system

$$A'x = b,$$

where A' is a matrix that only slightly differs from A. This will be the topic of the current section. Most prominently, we will show and explain an algorithm (see Algorithm 6, which we will denote by udiLU) to efficiently update the integral factorization diLU, that we have introduced in Section 4.3.2. This allows us to recover the factorization of a matrix

$$A' = A + yz^T, \tag{4.55}$$

where $y, z \in \mathcal{I}^n$, if we already have the integral factorization of the matrix A. This type of update amounts to an arbitrary change of rank-1 to the matrix A.

We will use this to implement the different basis matrix updates that were described in Section 3.4. If we study the different updates U_1-U_8 and U_{Z_1}-U_{Z_4}, we notice that all of them can be expressed as low-rank updates of the basis matrix. An update of type U_5 (see Figure 3.18a), for example, can be expressed as follows, $A' = A +$

$(c_{\text{new}} - c_{\text{old}})e_j^T$, where c_{new} is the column vector replacing the column vector c_{old} at position j, and e_j is the jth unit vector.

Even though some of the updates cannot be directly expressed as a rank-1 update, we can always compose a rank-r update from r successive rank-1 updates. The rank of the most complicated types of updates (U_{Z_2} and U_{Z_3}) is at most four. Also, note that it is possible to add or remove unit rows and columns from the back of the matrix, thus growing or shrinking the factorization. Doing this, and then adjusting the entries of the last row/column by the update procedure lets us implement the updates of types U_1-U_4.

The udiLU update procedure needs $\mathcal{O}(n^2)$ elementary operations. This is cheaper than computing the diLU factorization from scratch, which uses $\mathcal{O}(n^3)$ elementary operations.

There is one catch with this procedure, however. Our procedure does not allow for re-pivoting. The pivot order that was determined to factorize A has to be adopted for A' as well. This may lead to the undesirable situation that some element that was previously used as a pivot element becomes zero. In that case the update procedure fails. In Section 4.5.3, we are briefly going to sketch a heuristic that can be used to attempt recovery from such a *pivot failure*. It remains an open question, however, whether we can regenerate an arbitrary pivot order, and if yes, whether we can do that efficiently.

4.5.1. Related Work

There exists a considerable body of research dealing with different matrix factorization update procedures over the real numbers. To the best of our knowledge, in the realm of integral factorizations, our work presents the first result.

The closely related problem of updating the inverse of a matrix is surveyed by Hager [70]. The rank-1 update formula is known as the *Sherman-Morrison formula* [130, 129]. Independently, it was also given by Bartlett [14]. The formula for arbitrary rank changes is attributed to Woodbury [153]. Updating of the inverse in the context of the Schönherr's exact quadratic programming algorithm [128] is based on methods described by Gärtner [56], going back to Edmonds and Maurras [48]. Wessendorp [149] later added additional methods.

Similarly to the case of the inverse, methods have been developed to modify the LU factors of a matrix. Most notably, Bennet [18] and Gill, Golub, Murray, and Saunders [63]. The latter is not suitable for our purposes, because it relies on Householder reflections and Givens rotations to restore the factorization. The former, however, is the starting point for our own procedure.

In the context of sparse linear programming there has been a number of publications that deal with updating the basis matrix. Suhl and Suhl [137] describe an efficient method for restoring the triangular factors. They also describe earlier methods by Bartels and Golub [13], Forrest and Tomlin [54], and Reid [122]. Transcending the regular LU update as referred to in the previous paragraph, these methods try to reduce the number of necessary operations by exploiting the sparsity of the matrices involved. They do this by employing different permutation strategies in order to arrive at a favorable form for the update. Unfortunately, these methods are not easily adopted in the integral realm. As we have mentioned in the introduction to this chapter, it is not even clear how to apply any type of pivoting. Therefore, it also remains an open question whether we can employ any of these sparsity preserving techniques.

Let us go on to describing our update procedure.

4.5.2. Algorithm (udiLU)

For a detailed implementation of the algorithm udiLU see Algorithm 6. Here we will give a mathematical description of the principles behind it. The algorithm strives to recover the diLU factorization of $A' = A + yz^T$ if the factorization of $A \mapsto (L_\iota, U, L, U_\iota)$ is already known. Here $A, A' \in M_n(\mathcal{I})$ and $y, z \in \mathcal{I}^n$.

For simplicity let us assume that we will not have to deal with LU pivoting, that is, we will not have to concern ourselves with permuting any rows or columns of the matrices involved. For that we have to assume that the integral LU factorizations of A as well as A' exist, as defined in Section 4.3. In this case the restriction is more severe than before though. The update procedure described here *does not* lend itself to adaptive reordering. That is, even though there might be a pivot order for which the matrix A' has a valid integral factorization, this is not necessarily the case for an ordering that successfully factorized the matrix A. In a manner of speaking, the first matrix A to be factorized fixes the pivot order for all the updates that follow. This is undesirable for several reasons, and it can readily lead to a failed update attempt if one of the fixed pivot positions happens to become zero. This problem becomes especially pronounced for sparse matrices, which contain a lot of zero entries that are not suitable to serve as pivots. We will come back to this topic in Section 4.5.3. For now, let us assume that these problems are not of our concern.

The following equation will be the starting point of the description of the update process,

$$L_\iota A = U. \tag{4.56}$$

It it is clear that this holds if (L_ι, U) is a proper diLU factorization pair of A; see invariant (4.27). Considering the matrix $L_\iota A'$ we get the following,

$$L_\iota A' = L_\iota A + L_\iota y z^T = U + L_\iota y z^T.$$

Similar to equation (4.29) we can therefore look at our current problem as

$$(U + L_\iota y z^T) x = (U + y_\iota z^T) x = L_\iota b,$$

for some arbitrary vector $b \in R^n$ and solution vector x. Note the introduction of the simplifying variable $y_\iota := L_\iota y$. Exactly this kind of preprocessing is also done by Algorithm 6 in line 2. Now, the first important realization is that we could just apply the diLU process to that pair of matrices, $U + y_\iota z^T$ and L_ι. That is, we bring the former matrix into upper triangular form again by scaling and subtracting of rows. Instead of starting with a right-hand side I_n we shall start with the right-hand side L_ι instead. This will give the desired result, because ultimately the left-hand side will be reduced to upper triangular form again, and the right-hand side will be in lower triangular form. The latter is easy to see if we consider the changes made to the right-hand side and the fact that L_ι is lower triangular to start with. In particular, subtracting row i from row j, for $i < j$, does not introduce any nonzero elements above the diagonal. Also, scaling (with anything except 0) does not change the nonzero pattern of the matrix. This will give us (L_ι', U'), but the number of necessary computations is in the order of $\mathcal{O}(n^3)$, of course. In the following we will be concerned with explaining how to reduce the number of necessary computations to the order of $\mathcal{O}(n^2)$.

The important realization is that we have an *almost* upper triangular matrix $U + y_\iota z^T$ to start with. The property of being upper triangular is only disturbed by the rank-1 additive term $y_\iota z^T$. We can treat this case more efficiently than the general case. Let us see how. The algorithm goes row by row, treats each row exactly once, and in the end will have computed the factorization of A'. Again, for simplicity, we are only looking at the factorization pair (L_i, U) and omit the argument for the (L, U_i) part.

As was done in Section 4.3.2 we are going to claim an invariant that will hold at the beginning of each iteration of the outer loop of Algorithm 6 (lines 7 to 26),

$$A' = (L_i'^{(k)})^{-1} U'^{(k)}, \tag{4.57}$$

where $L_i'^{(k)}$ and $U'^{(k)}$ are defined in a similar fashion as well. They are the left and the right part of P, the matrix used as a working space by the algorithm,

$$
L_i'^{(k)} = \left(\begin{array}{c|c} L_{11}' & 0 \\ \hline L_{21}' & L_{22}' \end{array} \right) =
\left(
\begin{array}{ccccc|ccc}
q_1' & & & & & & & \\
* & q_2' & & & & & 0 & \\
\vdots & \ddots & \ddots & & & & & \\
* & \cdots & * & q_{k-1}' & & & & \\
\hline
* & & \cdots & & * & q_k & & \\
\vdots & & & & \vdots & \vdots & \ddots & \\
* & & \cdots & & * & * & \cdots & q_n
\end{array}
\right),
$$

and

$$U'^{(k)} = \left(\begin{array}{c|c} U'_{11} & U'_{12} \\ \hline 0 & U'_{22} \end{array} \right) = \left(\begin{array}{cccc|ccc} q'_2 & * & \cdots & * & * & \cdots & * \\ & q'_3 & \ddots & \vdots & & \vdots & & \vdots \\ & & \ddots & * & & & \\ & & & q'_k & * & \cdots & * \\ \hline & & & & * & \cdots & * \\ & & 0 & & \vdots & & \vdots \\ & & & & * & \cdots & * \end{array} \right).$$

Note that the structure has slightly changed. First of all, the elements below the diagonal in L'_{22} are nonzero, because we do not start with I_n. Second, the diagonal of L'_{22} is still initialized with the original diagonal elements of L_t. The symbol $*$ still stands for an unnamed nonzero entry. L'_{11}, U'_{11}, and U'_{12} are stable in the current iteration. Let us consider

$$U'_{22} = \left(\begin{array}{ccc} u'_{1,1} & \cdots & u'_{1,n-k+1} \\ \vdots & & \vdots \\ u'_{n-k+1,1} & \cdots & u'_{n-k+1,n-k+1} \end{array} \right).$$

again. There is s a second invariant that will help us to prove the correctness of Algorithm 6. It is that U'_{22} will always have the special property that it is the sum of an upper triangular matrix and a rank-1 matrix,

$$U'^{(k)}_{22} = U_{k..n,\,k..n} + y_{\imath k..n}(z^{(k)}_{k..n})^T. \tag{4.58}$$

Note that the first summand is a sub-matrix of the input matrix U.

In the second summand, the left side of the outer product is part of the vector y_i, which we have right from the start. Both of these only depend on k in the sense that we select a different range of entries in each iteration, but the entries themselves will always remain the same. Therefore, all the bookkeeping is left to the vector $z_{k..n}^{(k)}$, and as of now it is still a claim that such a vector exists. Let us have a look at the update rule for elements in U'_{22} that we should be using (recall equation (4.32)),

$$u'_{i,j} \leftarrow \frac{u'_{1,1}u'_{i,j} - u'_{i,1}u'_{1,j}}{q'_k}, \tag{4.59}$$

for all $k = 2,\ldots,n$ and $i,j = 2,\ldots,n-k+1$. Note that we have formulated the update rule for $k = 2,\ldots,n$ even though the outer loop of Algorithm 6 runs for $k = 1,\ldots,n$. The reason for this shift in starting indices is that, in udiLU, the first iteration is used to add $y_{i1}z^T$ to the first row, and then the z vector is updated to accommodate the second iteration. While we modify the elements of row k in iteration k, the update of succeeding rows is prepared in the current iteration. In other words, the update of row k was prepared in iteration $k-1$ by updating $z_j^{(k-1)}$. Turning back to the update equation, we can replace the variables by the original values according to invariant (4.58), e.g., $u'_{i,j} = u_{i,j} + y_{i}z_j^{(k)}$. We will drop the superscript k for a moment and only look at the numerator of the update rule. Then we get

$$u'_{1,1}(u_{i,j} + y_{i}z_j) - (u_{i,1} + y_{i}z_1)(u_{1,j} + y_{1}z_j) =$$

$$u'_{1,1}u_{i,j} + y_{i}(u_{1,1}z_j - u_{1,j}z_1) - u_{i,1}(u_{1,j} + y_{1}z_j) =$$

$$u'_{1,1}u_{i,j} + y_{i}(u_{1,1}z_j - u_{1,j}z_1). \tag{4.60}$$

The first equality follows by expanding the appropriate terms and

standard manipulation. The second equality follows because $u_{i,1} = 0$.
Recall that this denotes an original entry of U and for $i > 1$ this is 0
because U is upper triangular.

We have derived the update that is implemented in line 22 of Algorithm 6. The last term of (4.60) (in parentheses) corresponds to
the instructions of the algorithm. Note that k in the algorithm corresponds to 1 in our formula, and that the z of the formula is really
called z_1 in the algorithm. That is, in the formula the subscript designates accessing the first element, while in the algorithm it is just a
label. Unfortunately, this confusion of nomenclature is necessary, because in the algorithm we have to deal with two different variables z
because of the second factorization pair. Furthermore, in the algorithm, the variables u and u_r are used to store the original values of
the factorization.

The update of z in line 22 is then applied to the entries of the next
row to be updated *in the next iteration* of the loop, in lines 12 to 14,
which concludes the implementation of formula (4.60).

Unfortunately, we are not able to give the formal argument at this
point, why the divisions by q in the algorithm are integral.

4.5.3. Pivot Failure

We have already mentioned that the update procedure udiLU will fail
if any of the q'_k computed happens to be zero. Especially in the case
of sparse vectors this is likely to happen. Consider replacing a column
that only contains one nonzero entry with another that only contains
one nonzero entry, but in a different position. Since the pivot is linked
to a particular position in that column, we will encounter a zero pivot.

Unfortunately, we are not able to provide a systematic escape plan
in this situation. Changing the pivot order of an established integral

LU factorization is possible but computationally expensive (up to the point of having to do a complete re-factorization).

We did, however, implement a heuristic in the quadratic programming solver, which we call the *swap trick*. This heuristic is able to avoid pivot failures sometimes. To sketch it briefly, suppose we are trying to replace column c_{old} residing in position j by column c_{new}. Furthermore, assume that the position of the pivot in column j is i.

(i) Check *beforehand* whether c_{new} contains a nonzero element at position i. If yes, everything should be fine; abort this heuristic, carry out the rank-1 update using udiLU. If not, go to the next item.

(ii) Find a replacement column c_{rep} in the matrix that does contain a nonzero element in position i. Suppose the replacement's position is j' and its pivot is found at position i'. If c_{new} has a nonzero element at i' go to the next item, if not continue looking for another replacement column. If none is found, abort and report a pivot failure.

(iii) Carry out the swap. Using a rank-1 update, replace the column c_{rep} by c_{new}. Then, using another rank-1 update, replace c_{old} by c_{rep}. Swap j and j' in the column permutation of the factorization.

So, it may be possible to modify the permutation of a diLU factorization. In the general dense case, this is always possible. However, it is not efficient if we have to do several changes to the column permutation, and it is not adaptive within the udiLU update. In fact, we are *using* the udiLU update to achieve the modification. Therefore, one swap of the type described above needs $\mathcal{O}(n^2)$ operations. This is alright if we try to find a quick fix for one particular pivot failure, but if we have to establish a completely new permutation, this will be more expensive than re-factorization.

Finally, note that this applies to rows as well as to columns, of course. Also, it is possible to execute the swap trick for more complicated situations like the different basis matrix updates (see Section 3.4), but the procedure for these situations is more complicated because of the interdependence between the rows and columns to be exchanged. In the quadratic programming solver, we did implement provisions for the swap trick for some of the more complicated types of updates, with varying degree of success.

4.5.4. Complexity of udiLU

Let us derive the number of multiplications N_* of the udiLU update procedure first. We sum over the loops of Algorithm 6,

$$
\begin{aligned}
N_* &= \sum_{k=1}^{n} \left(\sum_{j=k}^{n+k} 2 + \sum_{j=k+1}^{n} 2 + \sum_{j=1}^{k-1} 2 + \sum_{j=k+1}^{n+k} 4 \right) \\
&= 2 \sum_{k=1}^{n} 4n \\
&= 8n^2.
\end{aligned}
$$

The number of divisions $N_{\div}$ and the number of subtractions N_- is the same, because each of inner loops contains an equal number of these. Therefore, we get

$$N_{\div} = N_- = \sum_{k=1}^{n} \left(\sum_{j=k}^{n+k} 1 + \sum_{j=k+1}^{n} 1 + \sum_{j=1}^{k-1} 1 + \sum_{j=k+1}^{n+k} 2 \right)$$

$$= 4n^2.$$

Computing the total number of operations we get

$$N = N_* + N_- + N_{\div} = 16n^2.$$

As we have claimed this is more efficient than computing the factorization from scratch.

Algorithm 6: Update-diLU (udiLU)

Input: $(L_\imath, U)$, $(L, U_\imath)$ the diLU factorization of A; $y, z \in \mathcal{I}^n$.
Output: $(L'_\imath, U')$, $(L', U'_\imath)$ the diLU factorization of $A + yz^T$.

1   $P \leftarrow [L + \text{triu}_1(U), \ L_\imath + \text{triu}_1(U_\imath)]$;
2   $y_1 \leftarrow L_\imath y$;
3   $z_1 \leftarrow [z; \mathbf{0}_n]$;
4   $y_2 \leftarrow U_\imath^T z^T$;
5   $z_2 \leftarrow [y; \mathbf{0}_n]$;
6   $q'_1 \leftarrow 1$;
7   **for** $k = 1$ **to** n **do**
8      $q \leftarrow P_{k, k+n}$;
9      $u \leftarrow P_{k, k}$;
10      $u_r \leftarrow P_{k, k+1..n+k}$;
11      $u_c \leftarrow [P_{k+1..n, k}; P_{1:k, n+k}]$;
12   **for** $j = k$ **to** $n + k$ **do**   ▷ Update U and $L_\imath$
13         $P_{k, j} \leftarrow (q'_k P_{k, j} + y_{1k} z_{1j}) \div q$;
14   **end**
15   **for** $j = k + 1$ **to** n **do**   ▷ Update L
16         $P_{j, k} \leftarrow (q'_k P_{j, k} + y_{2k} z_{2j}) \div q$;
17   **end**
18   **for** $j = 1$ **to** $k - 1$ **do**   ▷ Update $U_\imath$
19         $P_{j, n+k} \leftarrow (q'_k P_{j, n+k} + y_{2k} z_{2\,n+j}) \div q$;
20   **end**
21   **for** $j = k + 1$ **to** $n + k$ **do**   ▷ Update z_1 and z_2
22         $z_{1j} \leftarrow (u z_{1j} - z_{1k} u_{r\,j-k}) \div q$;
23         $z_{2j} \leftarrow (u z_{2j} - z_{2k} u_{c\,j-k}) \div q$;
24   **end**
25      $q'_{k+1} \leftarrow P_{k, k}$;
26   **end**
27   $L_\imath \leftarrow \text{tril}\big(P_{[*, n+1..2n]}\big)$;
28   $U \leftarrow \text{triu}\big(P_{[*, 1..n]}\big)$;
29   $L \leftarrow \text{tril}\big(P_{[*, 1..n]}\big)$;
30   $U_\imath \leftarrow \text{triu}\big(P_{[*, n+1..2n]}\big)$;

Implementation

After having discussed the theoretical aspects of the combination of the LU factorization and the simplex quadratic programming algorithm in the previous chapters, this chapter will be concerned with the practical implementation thereof. Many explanations in this chapter assume that the reader is familiar with the *C++ programming language* and the *Standard Template Library (STL)*. In particular, concepts like classes, templates, containers, and iterators should be known. See [136], for example, or any of the numerous online resources.

The quadratic programming solver is realized as a package of the *Computational Geometry Algorithms Library[18] (CGAL)* [49]. CGAL

---

[18] http://www.cgal.org/

is written in C++ and available under open source as well as commercial licenses. At the time of writing this thesis, the sparse features have not been added to the official distribution yet. Excerpts from the planned documentation are added in the Appendix A.6. In this chapter we are going to give a general overview and highlight some interesting implementation details.

Section 5.1 describes CGAL in general and gives an overview of the *Linear and Quadratic Programming* package in particular. Section 5.2 introduces the data structures used for sparse computations. Section 5.2.4 discusses the diLU factorization and the udiLU update. Finally, Section 5.3 discusses the linear and quadratic programming solver in more detail.

5.1. CGAL

The CGAL project was started in 1996 as a collaboration between several research institutions in Europe and Israel. Among these is also ETH Zürich. Quoting the introductory statement from its homepage is the most straightforward way of introducing it:

> The goal of the CGAL Open Source Project is to provide easy access to efficient and reliable geometric algorithms in the form of a C++ library. CGAL is used in various areas needing geometric computation, such as: computer graphics, scientific visualization, computer aided design and modeling, geographic information systems, molecular biology, medical imaging, robotics and motion planning, mesh generation, numerical methods...

CGAL offers a solid set of basic data types and data structures and a wide range of packages for more specialized problems that are mainly geometric in nature. The complete overview of packages can be found in the online manual[19]. To name a few examples, there are algorithms for computing convex hulls, Delaunay triangulations, Voronoi diagrams, arrangements, envelopes, meshes, and many other things. Up to today, there is quite a number of papers published containing CGAL in their title or abstract[20]. This is to say that, on the one hand, CGAL provides great tools to researchers to implement and test their ideas and algorithms. On the other hand, this is testimony to the fact that CGAL employs the newest and most efficient algorithms in many areas and is still actively being developed.

Last but not least, there are some packages that are not primarily geometric in nature, such as the *Linear and Quadratic Programming Solver* package. In the following sections we will focus on this package. Even though the relationship with geometry might not be immediately apparent, it was the main motivation for adding this package. Many geometric problems can be formulated in terms of linear or quadratic programs. In fact, two applications within CGAL already depend on the linear and quadratic programming solver; and there is another one being prepared. Those three packages are *minimum annulus*, *polytope distance*, and *extreme points*. The former two are already available in the distribution but the latter, at the time of writing this thesis, is still in the review process. Let us give a short explanation of those three applications, which are illustrated in Figure 5.1.

Minimum annulus is the problem of finding the (unique) pair of radii r and R, $r \leq R$, and center c, respectively, such that a given set

---

[19] http://www.cgal.org/Manual/latest/doc_html/cgal_manual/packages. html

[20] One example is [58]. We refrain from adding any other references.

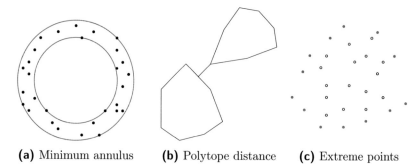

(a) Minimum annulus   **(b)** Polytope distance   **(c)** Extreme points

Figure 5.1.: Illustrations of the packages and features in CGAL that directly rely on the linear and quadratic programming solver.

of points is contained in the region between two concentric spheres that are placed at c with radii r and R, and such that $R^2 - r^2$ is minimized. See Figure 5.1a for an illustration. A formal description of this problem and the associated linear program is given in Section 5.5, where we also discuss the adaptions in the implementation that became necessary to comply with the sparse version of the solver.

The polytope distance problem asks to find the minimum distance between two polytopes, that are implicitly given by two sets of points respectively (the convex hull of the points defines the polytope); see Figure 5.1b. Similar as in the case of the minimum annulus problem, we will give the formal definition in a separate section, Section 5.4, where we discuss implementation details.

Finally, in the extreme points problem, one is again given a point set to start with; see Figure 5.1c. The task is to identify the points in this set that lie on the boundary of the convex hull of the whole point set. These points are called extreme points. Since this package did not need any particular adaptions in the context of the sparse version of the solver, we are going to defer the formal definition to the next chapter, Section 6.5.

The former two problems are also going to be featured in Chapter 6, where we discuss experimental results. Those two problems were introduced to CGAL by Schönherr [128]. The latter problem was implemented as a CGAL package by Helbling [71].

5.2. Sparse Tools

In this section we describe the implementation of a vector class `QP_sparse_vector` and a matrix class `QP_sparse_matrix`, that provide containers for matrix computations with a focus on exploiting sparsity, and their application in the framework of the CGAL quadratic programming solver.

5.2.1. Introduction

The main goal when dealing with sparse matrices is, of course, to avoid storing all those zero entries[21], while still keeping adequate accessibility and efficiency for the required operations. There are several different approaches of storing a sparse matrix.

Let us look at a simple example of a sparse matrix,

$$A = \begin{pmatrix} 4.5 & 0 & 3.2 & 0 \\ 3.1 & 2.9 & 0 & 0.9 \\ 0 & 1.7 & 3.0 & 0 \\ 3.5 & 0 & 0 & 1.0 \end{pmatrix}.$$

---

[21] More generally, sparse matrices can be based on an arbitrary default value, not necessarily zero. For the rest of this chapter, whenever we speak of a *nonzero entry*, we could substitute *non-default entry*.

A straight-forward idea for storing matrix A is to keep triplets of the form (`row_index`, `column_index`, `value`) for all the nonzero entries. The above matrix could therefore be stored as an array of triplets,

$$
\begin{aligned}
&[(0,0,4.5),\ (0,2,3.2),\ (1,0,3.1), \\
&\ (1,1,2.9),\ (1,3,0.9),\ (2,1,1.7), \\
&\ (2,2,3.0),\ (3,0,3.5),\ (3,3,1.0)]
\end{aligned}
\tag{5.2}
$$

Note that, as it is customary in the C++ language, we use zero based indexing here, i.e., the first element of a structure is assigned the index 0. This method of triplets, however, is somewhat cumbersome when implementing algorithms. We will therefore use STL containers to store the columns (or rows) of a matrix in a compressed format. This is not a novel idea (e.g. [113]). The above matrix can be stored as a set of four compressed row vectors,

$$
\begin{aligned}
[\ \ &[(0,4.5),\ (2,3.2)], \\
&[(0,3.1),\ (1,2.9),\ (3,0.9)], \\
&[(1,1.7),\ (2,3.0)], \\
&[(0,3.5),\ (3,1.0)] \qquad\qquad]
\end{aligned}
\tag{5.3}
$$

Each row in (5.3) now represents one row of the matrix. Note that the triples were reduced to pairs, because the row index of an entry is implicit in the storage format. This is the basic idea of our data structures. Chapter 2 of [37] discusses some variations in more detail.

To close, let us mention that there are open source implementations of sparse matrix tools available, such as *Eigen*[22], *SparseLib++*[23], or *TAUCS*[24], for example. It should be kept in mind, that some of these implementations might be more efficient than ours, and that the adoption of an external library must not be ruled out in the future. What made us decide for our own implementation, ultimately, are the special requirements of the integral LU factorization. These are *integral factorization, updating a factorization*, and *templatization for custom number types*. None of the libraries known to us features all these characteristics.

5.2.2. Vector Class

The data structure `QP_sparse_vector` that we are going to present will provide *single element access* in $\mathcal{O}(\log n_{\mathrm{nz}})$ and *iterator access* to all elements in $\mathcal{O}(n_{\mathrm{nz}})$, where n_{nz} is the number of nonzero elements of the vector. The former means that we can read, write, or modify an element of the vector, while the latter means that we sequentially visit all the elements of a vector and read, write, or modify them. What follows is the technical description.

The vector class `QP_sparse_vector` is implemented as a template, to be instantiated with an arbitrary number type `NT`. In addition to this number type, an instance of the vector class `QP_sparse_vector` is equipped with a size `n_` and a default entry `nt0_`. The default entry is used as the value of all those entries that are not specifically set to some other value. Usually `nt0_` is set to `NT(0)`, the zero element of the number type. The nonzero elements of the vector are stored in a

---

[22] http://eigen.tuxfamily.org/

[23] http://math.nist.gov/sparselib++/

[24] http://www.tau.ac.il/~stoledo/taucs/

map, pairing indices with values; see Listing 5.4.

```
template <typename NT>
class QP_sparse_vector {
  // Types
  typedef int key_type;
  typedef NT value_type;
  typedef typename std::map<key_type, NT> map_t;

  // data members
  int n_;
  NT nt0_;
  map_t entries_;
};
```

Listing 5.4: Basic class members of `QP_sparse_vector`.

Internally, the elements are stored in a `std::map`. For the client of the class, direct element access is provided by member functions such as the ones in Listing 5.5. However, note that – even though this is the conceptually easiest form of access – one should rely on iterator manipulations if possible and appropriate, for reasons described later.

```
// Getter & setter methods
const NT& get_entry(key_type n) const;
void set_entry(key_type n, NT val);
```

Listing 5.5: Getter and setter methods of `QP_sparse_vector`.

The problematic aspect of these functions is that they rely on the respective element manipulation functions of the `std::map`, i.e., their complexity is $\mathcal{O}(\log n_{\mathrm{nz}})$.

Conveniently, `std::map` supports bidirectional iteration through its elements in sorted order of its keys. We can utilize this to provide the following iterator interface to `QP_sparse_vector`; see Listing 5.6.

```
// Types
typedef typename map_t::iterator sparse_iterator_t;
// Accessors
sparse_iterator_t begin();
sparse_iterator_t end();
```

Listing 5.6: Iterator access to `QP_sparse_vector`.

As should be clear from those declarations we do nothing else than relay the `std::map::iterator` to the user of `QP_sparse_vector` through its own interface. According to the standard semantics, an instance it of type `QP_sparse_vector<NT>::sparse_iterator_t` can be incremented (`++it`), decremented (`--it`), and dereferenced (`*it`). The type of the latter is `std::pair<key_type, value_type>`. That is, we can access the index of an entry by `it->first` and the value by `it->second`.

If some operation reads or manipulates all the elements of a vector, it is more efficient to do this through iterators and *not* through the getter and setter methods. Let us assume that we have a vector $v \in \mathbb{N}^n$, and that we want to compute αv for some $\alpha \in \mathbb{N}$. Listing 5.7 illustrates this situation.

```
QP_sparse_vector<int> v;
int a;
...

// Manipulate entries through iterator access
for (QP_sparse_vector<int>::sparse_iterator_t it = v.
    begin(); it != v.end(); ++it) {
  it->second *= a;
}
```

```
// Inefficient way, using the setter method
for (QP_sparse_vector<int>::sparse_iterator_t it = v.
    begin(); it != v.end(); ++it) {
  v.set_entry(it->first, a * it->second);
}
```

Listing 5.7: Example of element access using `QP_sparse_vector`.

Both loops in the Listing 5.7 achieve the same thing. The second one is less efficient though, because for every nonzero entry one call to `set_entry` is issued. The complexity is therefore $\mathcal{O}(n_{\mathrm{nz}} \log n_{\mathrm{nz}})$. The first loop runs in linear time $\mathcal{O}(n_{\mathrm{nz}})$. There are situations, of course, where the methods `get_entry` and `set_entry` still make sense. If we only have to read or set one particular entry, for example. In that case, iterator access is less efficient, since we have to spend $\mathcal{O}(n_{\mathrm{nz}})$ only to get to the entry. This is why the implementation provides both methods.

For completeness, and because we will use it when describing the implementation of a sparse matrix, let us also mention the implementation of `operator[]`; see Listing 5.8.

```
template <typename NT>
const NT& operator[] (const key_type& n) const {
  return get_entry(n);
}
```

Listing 5.8: Implementation of `operator[]` for `QP_sparse_vector<NT>`.

Note that we do not provide a non-const version of `operator[]`. For `std::map`, which is the underlying data structure here after all, the C++ standard [1] defines in Section 23 that a reference `NT&` should be returned in that case. This means, if we tried to access an element whose key does not exist yet, we have to insert `std::pair<key_type, NT>(n, nt0_)` into the vector. In our case this behavior is *not* desired, because it fills up the vector with de facto zero entries for every read access to a zero element.

5.2.3. Matrix Class

Building on the sparse vectors from the preceding section, the sparse matrix data structure `QP_sparse_matrix` provides access to a single element in $\mathcal{O}(\log n_{\mathrm{nz}}^{\mathrm{r}_i})$, and iteration over its rows or columns in $\mathcal{O}(n_{\mathrm{nz}}^{\mathrm{r}_i})$ and $\mathcal{O}(n_{\mathrm{nz}}^{\mathrm{c}_j})$ respectively. Here $n_{\mathrm{nz}}^{\mathrm{r}_i}$ is the number of nonzero entries in row i, and $n_{\mathrm{nz}}^{\mathrm{c}_j}$ is the number of nonzero entries in column j.

The matrix class `QP_sparse_matrix` is used to keep the basis matrix in the quadratic programming solver. It is the main data structure upon which the LU factorization builds.

```
template <typename NT>
class QP_sparse_matrix {
  // data members
  int m_; // row dimension
  int n_; // column dimension
  NT nt0_;
  std::vector<QP_sparse_vector<int> > rows_, columns_;
  std::vector<NT> data_;
  int next_index_;
};
```

Listing 5.9: Basic data members of `QP_sparse_matrix`.

Listing 5.9 requires some explanations. First of all – as in the case of sparse vectors – there are data elements that indicate the size of the matrix, such as `m_` and `n_` for the number of rows and columns respectively. Also, there is a default entry `nt0_`, which will usually be zero.

The remaining data elements are used to store the nonzero elements of the matrix. The main idea is that we separate the logical index structure, which is stored in the two variables `rows_` and `columns_`, from the actual data elements, which are stored in an array `data_`. See Figure 5.10 for an illustration of the index structure. As you can see in the declaration of `rows_` and `columns_` these two arrays store sparse vectors containing integer variables. These are the actual indices into `data_`. The reason for this setup is that we want to be able to do manipulations without having to drag around numerous instances of `NT`. Only when we are sure about accessing an entry do we use the integer index to actually access it. The data member `next_index_` indicates the next free cell in the array `data_`.

There is one redundant element in the implementation of `QP_sparse_matrix`. That is, it has a structure that keeps track of the rows *and* the columns of the matrix. In fact, each of the arrays `rows_` or `columns_` is sufficient to reconstruct the *whole* matrix. The reason why we have both of these structures is that we want to be able to efficiently iterate over the rows as well as the columns of the matrix. This is a consequence of the fact that we want to be able to compute the LU factorization of some matrix A as well as of its transpose A^T (see Section 4.3). Unfortunately though, we do not have any control over the template number type `NT`, in general. In particular, `QP_sparse_matrix` is going to be instantiated with an exact number type in the course of the LU factorization. Hence it seems reasonable to minimize the number of instances of and operations on variables of type `NT`. We do not know how `NT` is implemented, and the numbers during the LU factorization may grow large.

For these reasons the entries of the matrix are stored in the one di-

mensional array `data_`. In `rows_` and `columns_` we only store indices
into that array. To be consistent, it must always hold that

$$\texttt{rows\_[i][j] == columns\_[j][i],}$$

for all $0 \leq \texttt{i} < \texttt{m\_}$ and $0 \leq \texttt{j} < \texttt{n\_}$. This is an invariant of
`QP_sparse_matrix`. Note that, in the above formulation, the first ap-
plication of `operator[]` is simple index arithmetic on a `std::vector`,
while the second application is a call to `get_entry` as described in
Section 5.2.2.

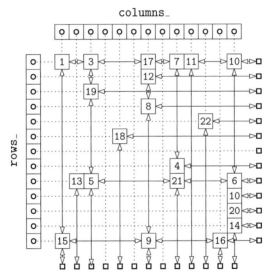

Figure 5.10.: Shows an example of the logical indexing structure of the
class `QP_sparse_matrix`. The cells contain integer indices
that point to an array (not shown) of values. The dotted lines
indicate the individual rows and columns. The solid arrows
represent iterator access for one particular row or column,
and the small boxes represent the past-the-end iterators. Note
that not every index needs to be present, as index 2 is missing,
for example.

Irrespective of these considerations, `QP_sparse_matrix` offers a similar set of operations and access routines as `QP_sparse_vector`. The straight-forward single element access is provided by methods like the ones in Listing 5.11.

```
// Getter & setter
const NT& get_entry(int m, int n) const;
bool set_entry(int m, int n, const NT& val);
```

Listing 5.11: Getter and setter methods of `QP_sparse_matrix`.

As in Section 5.2.2, where we discussed access to a sparse vector, we also want iterator access which is more tuned towards sparse inputs. This is provided by a set of `begin` and `end` functions to obtain iterators for single rows and columns. See Listing 5.12 for an example. There is a distinction between `index_iterator` and `value_iterator`. The former is normally not needed by a user of `QP_sparse_matrix`, because it iterates over the internal indices into the array `data_`, while the latter is the standard way of iterating over the entries of the matrix.

```
// Types
typedef typename QP_sparse_vector<int>::sparse_iterator_t
    index_iterator;
typedef typename boost::transform_iterator<Index_to_NT,
    index_iterator> value_iterator;
// Iterator access
value_iterator begin_column_value(int i);
value_iterator end_column_value(int i);
```

Listing 5.12: Iterator access to `QP_sparse_matrix`.

The functor `Index_to_NT` (see Listing 5.13) is used to, roughly speaking, turn an instance of `index_iterator` into an instance of `value_iterator`. Note that this returns pointers to the entries, that is, if `it` is of type `value_iterator`, `it->first` provides an index (of the matrix) while `*it->second` provides the value. Note that the second element of `result_type` is a pointer to `NT`. Therefore, we have to dereference once more to get to the value.

```
struct Index_to_NT {
    typedef typename std::pair<int,int> input_type;
    typedef typename std::pair<int, NT*> result_type;

    Index_to_NT(): data_pointer_(0) {}
    Index_to_NT(std::vector<NT>* p): data_pointer_(p) {}

    result_type operator() (const input_type& p) const {
      return result_type(p.first, &(data_pointer_->at(p.
         second)));
    }

  private:
    std::vector<NT>* data_pointer_;
};
```

Listing 5.13: Helper object that provides pointers to the entries of the matrix when given an index.

5.2.4. LU Factorization

The integral LU factorization is implemented as a stand-alone class but access to it is greatly tuned to the needs of the quadratic programming solver. It is parameterized by a pointer to the `Matrix_provider` class that is required to provide the matrix to be factorized on demand.

```
template <..., typename Matrix_Provider>
class QP_LU_factorization {
  public:
    template < typename InIt, typename OutIt >
    void  solve(InIt v_l_it, InIt v_x_it,
                OutIt y_l_it,  OutIt y_x_it);
    bool rank_1_update(QP_sparse_vector<ET> y,
                       QP_sparse_vector<ET> z);

  private:
    void compute_factorization();

    template < class InIt, class OutIt>  // QP case
    void  solve_QP( InIt v_l_it, InIt v_x_it,
                   OutIt y_l_it, OutIt y_x_it);
    template < class InIt, class OutIt>  // LP case
    void  solve_LP( InIt v_l_it, InIt v_x_it,
                   OutIt y_l_it, OutIt y_x_it);
};
```

Listing 5.14: Class implementation of the integral factorization.

Factorization is usually not initiated manually. Instead it is triggered by the access method `solve`. Once the user requests to solve an equation system, we check whether the matrix had previously been factorized. If the factorization does not exist, it is lazily evaluated on demand. Note that the parameters of `solve` are reminiscent of equation (3.8). The input and the output are subdivided into the λ-part and the x-part[25]. This is done because we handle the LP case differently from the QP case. In the former situation we only store the matrix A, because M_B takes the simplified form of equation

---
[25] See Section 3.3 for an explanation of the λ- and x-part.

(3.14). We can see this distinction implemented in the two private methods `solve_QP` and `solve_LP` that are used internally to compute the solution. In the private section we further notice the method `compute_factorization` that is used internally to trigger factorization from scratch if need be.

Last, we have included the method `rank_1_update`, which is accessible publicly. It can be used by the caller to modify the internal LU representation of the matrix stored (assuming that the factorization already exists). The parameters `y` and `z` are according to equation (4.55). Calls to this method are used to implement all the different updates as described in Section 3.4.

5.3. Quadratic Programming Solver

The interface of the quadratic programming solver is hardly changed by our implementation. Most important, the new `LU` version of the solver is backward compatible with the previous code, except for one detail. Previously, it was possible to retrieve dense iterators for the matrices A and D[26] from the program, i.e., by calling `get_a()`, which returns an iterator to the matrix A. This has been replaced by sparse access, i.e., `get_a_sparse()`, which also returns an iterator to the matrix A but one that has an (`index,value`) structure. It is still possible to construct the program from dense iterators. However, this feature comes with certain restrictions concerning efficiency. We will discuss some of these issues in the following Sections 5.4 and 5.5, and in the chapter about the experimental results, Chapter 6.

Other than this, we refrain from describing the technical interface

---

[26] Recall the matrices A and D from the definition of a quadratic program (1.1).

at this point and refer to the CGAL documentation[27]. In Appendix
A.6 we include a few pages from the updated documentation. This
outlines the modified *concept* `QuadraticProgram` that defines the interface for a quadratic programming instance. All the previous implementations (called *models*) of this concept are still valid, and there
is a new model `Quadratic_program_from_sparse_iterators`.

5.4. Polytope Distance

This problem is a classical problem of computational geometry. A solution is implemented as a package[28] in CGAL and described in Schönherr's PhD thesis [128]. It fully depends on the quadratic programming solver. We will restate the mathematical definition of the problem and explain the approach by which it is solved, for consistency.
For an illustration see Figure 5.1b. The main point of this section,
however, is to describe how the LU version of the quadratic programming solver can be adapted to work with the sparse iterator input
interface. Experimental results about this approach can be found in
Section 6.3.

Problem 5.15 (Polytope distance problem). *Given two sets of points*
$P, Q \subset \mathbb{R}^d$, *determine the minimum distance between the convex hulls*
of the two sets.

The definitions of convex hull of a point set and distance between two
point sets are as follows.

---

[27] http://www.cgal.org/Manual/latest/doc_html/cgal_manual/packages.
html
[28] http://www.cgal.org/Manual/latest/doc_html/cgal_manual/Polytope_
distance_d/Chapter_main.html

Definition 5.16. *The* convex hull, $\mathrm{conv}(P)$, *of a finite point set P is defined to be the set of all convex combinations of P. Formally,*

$$\mathrm{conv}(P) := \left\{ \sum_{p \in P'} \lambda_p p, \text{ for } P' \subseteq P, \ \lambda_p \geq 0, \text{ and } \sum_{p \in P'} \lambda_p = 1 \right\}.$$

Definition 5.17. *For two points sets $P, Q \subset \mathbb{R}^d$ we define their* distance $\mathrm{d}(P, Q)$ *as*

$$\mathrm{d}(P, Q) := \min_{p,q} \|p - q\|_d,$$

where $p \in \mathrm{conv}\, P$ and $q \in \mathrm{conv}\, Q$.

Let the point sets consist of the following points, $P = \{p_1, \ldots, p_r\}$ and $Q = q_{r+1}, \ldots, q_n$. Let $p \in \mathrm{conv}(P)$ and $q \in \mathrm{conv}(Q)$ be the two points that attain the minimum distance between P and Q. Furthermore, let v and w be two nonnegative d-vectors such that $v - w = p - q$. The solution of Problem 5.15 can be formulated as the following quadratic program,

$$
\begin{aligned}
\text{(PD)} \qquad \min \quad & (v - w)^T(v - w) \\
\text{s.t.} \quad & v - w - \sum_{i=1}^{r} x_i p_i + \sum_{i=r+1}^{n} x_i q_i = 0 \\
& \sum_{i=1}^{r} x_i = 1 \\
& \sum_{i=r+1}^{n} x_i = 1 \\
& x_i \geq 0, \quad 1 \leq i \leq n \\
& v_i, w_i \geq 0, \quad 1 \leq i \leq d.
\end{aligned}
\qquad (5.18)
$$

Here, the solution variables are all the entries of v and w and also

x_i for all $1 \leq i \leq n$. The first constraint is a vector constraint that formulates p and q as linear combinations of the points in P and Q, and assures that $v-w = p-q$. The second and third constraints make sure that the linear combinations are convex combinations, and that the resulting points are therefore contained within the convex hulls of P and Q. The objective function is simply the squared distance $\|v - w\|_2^2 = \|p - q\|_2^2$.

To summarize, the matrices A and D have the following form. We assume that the ordering of the variables lists all unknowns of v, w, and then all x_i,

$$
A = \begin{pmatrix} I_d & -I_d & -P & Q \\ 0 & .0 & 1 & 0 \\ 0 & 0 & 0 & 1 \end{pmatrix}, \tag{5.19}
$$

where P and Q are the points of the two inputs sets arranged in columns. I_d is the identity matrix of order d, and the last two rows in (5.19) are single rows. The four 0 entries in the lower left part are therefore row vectors containing d zero entries. The four entries in the lower right part are row vectors containing n entries each. The dimension of A is $(d + 2) \times (n + 2d)$. This matrix is not particularly sparse. The situation is different for D,

$$
D = \begin{pmatrix} I_d & -I_d & 0 \\ -I_d & I_d & 0 \\ 0 & 0 & 0 \end{pmatrix}. \tag{5.20}
$$

Considering that none of the variables x_i appear in D and that usually

$n \gg d$, this matrix is extremely sparse. The 0 entry in the lower right part is a sub-matrix of dimension $n \times n$ (which implicitly defines the dimensions of the other 0 entries as well). The dimension of the whole matrix D is $(n + 2d) \times (n + 2d)$.

To comply with the sparse iterator input interface (see Section 5.3), the matrices A and D are implemented as follows. We only give the description for D. The data structures for A are realized analogously.

```cpp
// iterator for a fixed row of D
template <typename NT_>
class D_sparse_column_iterator :
  public boost::iterator_facade<...>
{
  // typedefs
  ...

  // constructor
  explicit D_sparse_column_iterator(int j, int d, int i)
    : j_(j), i_(i), d_(d) {
      adjust_indices();
  }

private:
  // Used to skip zero elements
  void adjust_indices() const;

  // required by iterator_facade
  void increment();
  void decrement();
  bool equal(const Derived & other) const;
  value_type& dereference() const;

private:
  int j_; // column number
```

```
    int i_; // row number
    int d_; // dimension
    mutable value_pair tmp_;
};
```

Listing 5.21: Refinement of `boost::iterator_facade`, defining column
iterators for D.

Iterators for one specific column of D can be realized by refin-
ing `boost::iterator_facade`[29]. In Listing 5.21 we left out all
the `typedefs` and many other requirements of `iterator_facade` for
brevity (indicated by "..."). Note that the type `value_pair` is simply `std::pair<int, NT>`. The data members are as follows: `i_`, `j_`,
and `d_` are used to store the current position (row) of the iterator,
the column of D that it represents, and the total row dimension of D,
respectively. The member `tmp_` is of type `value_pair` and is used to
store the return value, in case the iterator will be dereferenced. This
is necessary, because iterator semantics require that we return a reference. The member `tmp_` is declared `mutable`, because comparison of
two iterators should yield the same result independent of whether or
not any of the iterators had been dereferenced earlier (which changes
`tmp_`). It is important to note that the references obtained from the
iterator constructed *must not* be kept for future use. This is a downside of this approach, but the advantage of being able to generate the
matrix elements on demand (as opposed to having to store the whole
matrix) is of greater interest.

The idea is as follows. The `iterator_facade` interface requires
us to implement the functions `increment`, `decrement`, `equal`, and
`dereference`. Essentially these provisions are sufficient to let

---

[29] http://www.boost.org/doc/libs/1_50_0/libs/iterator/doc/iterator_
facade.html

`D_sparse_column_iterator` appear as any other standard C++ iterator. The function `adjust_indices` was added to move the iterator to the next nonzero position. For example, in the implementation of `increment`, we advance `i_` by one, and then call `adjust_indices`. It is an invariant of the iterator that it should always point at a nonzero entry (or the past-the-end position). So, after any manipulation of `i_` we have to adjust the indices.

This iterator adaption is really the core of the implementation changes of the polytope distance problem. It provides a sparse iterator that returns pairs of the form (`index, value`) for D. This needs to be further embedded in a type `D_sparse_column`, from which to derive the iterator, and then ultimately in types `D_sparse_matrix_iterator` and `D_sparse_matrix` to provide access to the whole matrix.

```
template <typename NT_>
class D_sparse_column {

  ...

  D_sparse_column_iterator<NT> begin() {
    return D_sparse_column_iterator<NT>(j_, d_, 0);
  }

  D_sparse_column_iterator<NT> end() {
    return D_sparse_column_iterator<NT>(j_, d_, 2*d_);
  }

  ...

private:
  int j_; // col number
  int d_; // dimension
};
```

Listing 5.22: Class representing the columns of D.

```
template <typename NT_>
class D_sparse_matrix: public std::unary_function <int,
    D_sparse_column<NT_> >;


template <typename NT_>
struct D_sparse_matrix_iterator {
  typedef typename boost::transform_iterator<
      D_sparse_matrix<NT_>,
      typename boost::counting_iterator<int> > Type;
};
```

Listing 5.23: Classes representing the whole matrix D.

We have included the skeleton of the additional data structures necessary to represent D in Listing 5.22 and Listing 5.23, hoping that they are self-explanatory to the interested reader.

5.5. Minimum Annulus

As for the problem of the previous section, a solution to the *minimum annulus* problem is implemented in CGAL and it was described in Schönherr's thesis [128]. Here, we will give the formal definition and some implementation details. For an illustration of the minimum annulus problem see Figure 5.1a.

Definition 5.24. *An* annulus *is the region between two concentric spheres in $\mathbb{R}^d$ with center c and radii r and R, where $r \leq R$. Formally, it is the set*

$$\{p \in \mathbb{R}^d \mid r^2 \leq (p-c)^T(p-c) \leq R^2\}.$$

The minimum annulus problem can be formulated as follows.

Problem 5.25 (Minimum annulus problem). *Given a finite point set* $P \subset \mathbb{R}^d$, *determine the annulus that contains* P *and minimizes* $R^2 - r^2$.

Making the substitution $u := R^2 - c^T c$ and $v := r^2 - c^T c$, we can formulate the problem as the following primal linear program,

$$
\begin{aligned}
\text{(MA)} \quad \max \quad & v - u \\
\text{s.t.} \quad & p^T p - 2p^T c \geq v, \quad \forall p \in P \qquad (5.26) \\
& p^T p - 2p^T c \leq u, \quad \forall p \in P.
\end{aligned}
$$

The reader is invited to make the substitutions for u and v to see the evident formulation of the problem. Introducing a dual variable x_p for every constraint of the first type, a variable y_p for every constraint of the second type, and some transformations, we obtain the following dual program,

$$
\begin{aligned}
\text{(MA')} \quad \min \quad & \sum_{p \in P} x_p p^T p - \sum_{p \in P} y_p p^T p \\
\text{s.t.} \quad & 2 \sum_{p \in P} x_p p - 2 \sum_{p \in P} y_p p = 0 \\
& \sum_{p \in P} x_p = 1 \qquad (5.27) \\
& \sum_{p \in P} y_p = 1 \\
& x_p, y_p \geq 0, \quad \forall p \in P.
\end{aligned}
$$

This is the problem that is solved in the CGAL implementation. The matrix A for MA' looks as follows,

$$A = \begin{pmatrix} 2P & -2P \\ 1 & 0 \\ 0 & 1 \end{pmatrix}, \tag{5.28}$$

where $P = \{p_1, \ldots, p_n\}$ contains the points as columns, and the entries in the lower part of A are row vectors of n entries each. The dimension of A is $(d+2) \times 2n$. There is no matrix D since the problem is linear.

The implementation of this package was not changed to a sparse input interface yet. We have left it like this for lack of time, but also to demonstrate that the LU version of the solver is backward compatible with the old interface. Dense input iterators are allowed. The test results in Section 6.4, however, show that this is not a desirable situation, because it leads to slow running times. If efficiency is an issue, a sparse input interface should be used.

Let us briefly describe how the adaption of a dense input interface works in the sparse version of the solver. To differentiate between sparse and dense input interfaces, all incoming iterators are funnelled through an adaptor class. This adaptor provides access to a sparse version of the input, whichever interface is used for the input. Listing 5.29 shows the skeleton of this adaptor class.

```
template < typename Q_, typename Is_sparse_ >
struct Sparse_iterator_adaptor {
   typedef Is_sparse_ Is_sparse;
};

// sparse specialization
template < typename Q_>
```

```
struct Sparse_iterator_adaptor<Q_, CGAL::Tag_true>
{
  typedef Q_ Quadratic_program;
  typedef Q_::A_sparse_iterator A_sparse_iterator;
  typedef Q_::A_sparse_column_iterator
     A_sparse_column_iterator;
  typedef true Is_sparse;

  A_sparse_iterator get_a_sparse(const Q_& qp) {
    return qp.get_a_sparse();
  }
};

// dense specialization
template <typename Q_>
struct Sparse_iterator_adaptor<Q_, CGAL::Tag_false>
{
  typedef Q_::A_iterator A_it;
  typedef ... A_sparse_iterator;
  typedef ... A_sparse_column_iterator;
  typedef false Is_sparse;

  A_sparse_iterator get_a_sparse(const Q_& qp) {
    return A_sparse_iterator(qp.get_a(), ...);
  }
};
```

Listing 5.29: Iterator adaptor class with two specializations for the sparse and the dense case respectively.

We only want to get the basic idea of this approach across. Therefore, the code in Listing 5.29 is highly simplified and only mentioning the matrix A. In the actual implementation there are provisions for D as well. The main idea is that we have the template adaptor class `Sparse_iterator_adaptor` that is parameterized by the type of the quadratic program that it is meant to operate with and a Boolean tag indicating whether this program is dense or sparse. The tag is provided by the quadratic program itself. Then we define two specializations of that class which both define the function `get_a_sparse`. This is to be called using a quadratic program (of the appropriate type) as parameter. Depending on the type, the template switch guarantees that the right function is compiled into the code.

The *sparse specialization* does nothing else than relay the types and iterators of the quadratic program, because it knows that those are sparse already. In the *dense specialization* more work is necessary. We have to extract the iterator type `A_it` from the quadratic program, and then wrap it and its columns in appropriate constructions `A_sparse_iterator` and `A_sparse_column_iterator`, so it will behave like a sparse matrix. This is not completely trivial, and requires further class definitions. We have omitted mention of those custom types, which is indicated by ellipses in the listing. Ultimately, what is important is that the function `get_a_sparse` returns the dense iterator wrapped in the newly defined type `A_sparse_iterator`. The advantage of this approach is that we can always assume that we get access to a sparse iterator, simply by calling `get_a_sparse`.

This concludes our description. For further details please see the CGAL documentation or the source code of the implementation.

I love fools' experiments.
I am always making them.

Charles Darwin

6

Experimental Results

We are going to discuss different test scenarios in the following sections. There are a multitude of quadratic programming solvers available; for example CPLEX[30], Gurobi[31], FinMath[32], SNOPT[33], KNITRO[34], MOSEK[35] [7, 8, 6], to name a few of the best known commercial ones. There is also integration of quadratic programming solvers

---

[30] http://www.ibm.com/software/integration/optimization/cplex-optimizer/
[31] http://www.gurobi.com/products/gurobi-optimizer/
[32] https://rtmath.net/products/finmath/
[33] http://www.sbsi-sol-optimize.com/asp/sol_products_snopt_desc.htm
[34] http://www.ziena.com/knitro.htm
[35] http://mosek.com/products/mosek/

into well-known mathematical computation environments like MAT-LAB[36], Maple[37], and Mathematica[38], through optimization packages or specialized plug-in packages. Access to these software suites is commercially restricted. There are a few open source packages, such as the quadratic programming solver of the GNU Octave project[39], or the OpenOpt initiative[40].

To the best of our knowledge, none of these solvers employ exact arithmetic. Therefore, it is unfair to compare the CGAL solver to any of them in terms of applicable range of parameters, especially the ones implementing some fast interior point methods. For this reason – but also because the main contribution of this work was to enable the current CGAL implementation exploit sparse inputs – the main focus of the comparison will be between two versions of CGAL. Throughout this chapter we will denote the old version of the solver, which uses direct inversion, by `INV`. By `LU`, on the other hand, we will denote the version that employs LU factorization techniques.

The following sections will explain in more detail what the different test cases and findings are. Section 6.1 will introduce test instances from the Netlib repository. The following section, Section 6.2, investigates randomly generated instances. Finally, Sections 6.3, 6.4, and 6.5 are going to discuss the geometrical problems within the CGAL library that rely on quadratic (or linear) programming formulations. All tests were conducted on an Apple MacBook Pro, Intel Core 2 Duo processor, 2.8 GHz, with 4 GB RAM.

---

[36] http://www.mathworks.ch/products/matlab/

[37] http://www.maplesoft.com/products/Maple/

[38] http://www.wolfram.com/mathematica/

[39] http://www.gnu.org/software/octave/

[40] http://openopt.org/

6.1. Netlib Cases

The Netlib repository[41] is an online repository of freely available programs, documents, and databases of interest to the numerical and scientific computing community. It is maintained by AT&T Labs[42], the University of Tennessee[43], and Oak Ridge National Laboratory[44]. It is of special interest to us, because it features a collection of sparse linear programs in the MPS format[45]. Many of these programs are to large to be handled by the CGAL solver. Some of the inexact, commercial solvers mentioned earlier can do that, but we will restrict ourselves to the following cases and compare the previous version of the solver (INV) with the sparse one (LU).

There are several exact linear programming solvers that we could also use to compare with. Some that are freely available are CDD[46], LRS[47], QSopt-Exact[48], and EXLP[49]. We only include comparisons with the last one, because it conveniently supports the MPS interface, as the CGAL solver does too.

In Table 6.1 we outline the specifications of the cases we investigated. All of these cases are extremely sparse, except for FIT1D and FIT2D. We include these for checking the overhead incurred by the sparse handling. In Table 6.2 you will find the running times of the

---

[41] http://www.netlib.org/

[42] http://www.bell-labs.com/

[43] http://www.utk.edu/

[44] http://www.ornl.gov/

[45] Described in the CGAL documentation http://www.cgal.org/Manual/latest/doc_html/cgal_manual/QP_solver_ref/Concept_MPSFormat.html

[46] http://www.ifor.math.ethz.ch/~fukuda/cdd_home/

[47] http://cgm.cs.mcgill.ca/~avis/C/lrs.html

[48] http://www.dii.uchile.cl/~daespino/QSoptExact_doc/main.html

[49] http://members.jcom.home.ne.jp/masashi777/exlp.html

selected Netlib cases. All tests were run with GMP number types. In particular, in CGAL this is the wrapper class gmpzf.

Name	m	n	d	opt
AFIRO	28	32	0.098	-4.65E+02
CAPRI	272	354	0.019	2.69E+03
FIT1D	25	1026	0.563	-9.15E+03
FIT2D	26	10500	0.506	-6.85E+04
FORPLAN	162	421	0.072	-6.64E+02
SCSD1	78	760	0.053	8.67E+00
SHARE1B	118	225	0.045	-7.66E+04
STOCFOR1	118	111	0.036	-4.11E+04

Table 6.1.: Summary of the tested Netlib cases. The columns m and n indicate the number of (in)equalities and variables of the linear program, as usual. The values $0 \leq d \leq 1$ indicate the *density* of the problem, that is the fraction of nonzero entries in A. The final column gives the optimal value of the problem.

Problem	LU	INV	LU/INV	EXPL
AFIRO	0.006094	0.012834	0.47	0.167
CAPRI	72.5459	92.151	0.79	0.365
FIT1D	24.3692	29.1307	0.84	0.610
FIT2D	1573.45	2595.14	0.61	26.720
FORPLAN	12.8461	26.0381	0.49	0.620
SCSD1	12.3181	38.9794	0.32	0.118
SHARE1B	15.1337	16.8949	0.90	0.205
STOCFOR1	0.683486	2.05473	0.34	0.059

Table 6.2.: A table of the running times incurred by solving the Netlib cases. The results are given in seconds, except for the fourth column (LU/INV), which is the speed-up factor of the sparse variant. We measured the complete execution time, including problem input and actual solve time. For the two CGAL variants, the solver routine solve_linear_program was used.

The first thing we will briefly address is that the running times of

EXPL are a lot better than the CGAL running times; up to one or two orders of magnitude. Unfortunately, we do not have a conclusive explanation for that phenomenon, as our main concern lay on the intrinsic improvement of and the comparison with the previous CGAL solver. One possible explanations for the strikingly good performance of EXLP is the fact that it only deals with linear programs. Even though the solver routine `solve_linear_program` tries to avoid superfluous computations and function calls associated with the solution process of a proper quadratic program, the whole procedure is of course embedded in a more complicated framework. EXPL is light-weight and *does* use divisions over a field. The LU factorization it uses is comparable to ours and also uses the Markowitz heuristic to find the next LU pivot element. Since the factorization is driven by an eta-file, the size of the numbers in the computation is not systematically bounded. That possible penalty cannot be prohibitive, however, as the running time proves. On the other hand, the reasons for the comparatively good performance of EXPL might also be extraneous and have something to do with the number type used, for example. Even though both solvers use GMP, the CGAL type has a wrapper class around it that might incur considerable overhead. We propose to investigate this behavior further.

Coming back to the comparison between LU and INV, we can see that the speed-up factor – in some cases – is as good as 0.32, but in other cases the running times are almost the same. We cannot fully explain this but influencing factors are the structure of the basis matrix, the fill-in that is generated, and the specific set of basis updates that have to be done. Recall that – in the case of the LU factorization – the solver suffers from possible pivot failures (see Section 4.5.3). Depending on which updates have to be done, and which of these fail, the running time is affected more or less. Let us give a few more experimental pointers as to what happens in the above Netlib cases in the following section.

6.2. Random Cases

In this section we present results that are obtained by considering different random settings. Some of these settings are implemented in the test suite of the quadratic programming solver in CGAL, and have proved successful in finding errors in the code. By visiting a large number of small cases it is possible to generate many rare constellations.

We have subdivided the results in two subsections, 6.2.1 and 6.2.2, which are not listed in the table of contents. These two sections discuss completely randomized instances of varying size.

6.2.1. Small Cases

In this setting we look at small cases of 2 variables and 2 constraints each. Everything else is randomized, the entries of the matrices A, D, and also the entries of the vectors c and b are chosen as integer numbers uniformly at random in the interval [-10, 10]. The relations of the constraints in $Ax \gtreqless b$ and the upper and lower bounds for the variables are chosen randomly as well.

Once an instance has been generated, it is solved with all the different solver routines of the quadratic programming solver. That is, the instance is considered as a linear program as well as a proper quadratic program. Also, we distinguish between nonnegative and instances not having standard bounds. This gives four different solution routines. If any of the input data is not relevant for the chosen solution method, it is ignored, e.g., the matrix D in case we solve the instance as an linear program. In Figure 6.3 we include a statistic of how the solution of these problems turn out to be. In this small setting there is not much difference between any of the solution methods. For the results included here, we generated a total number of $3 \cdot 10^5$ cases.

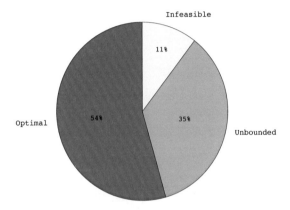

Figure 6.3.: Small random cases as described in Section 6.2.1 are solved. The distribution of *optimal, unbounded* and *infeasible* cases is shown. The plot represents an accumulation of $3 \cdot 10^5$ cases in total.

We have solved half of the cases with the LU variant, and half of the cases with the old INV variant. Table 6.4 summarizes results we obtained with regards to the running time. We have divided the cases into batches of 50000 each. This way we can screen the stability of our measurements, i.e., ensure that different cases do not produce wildly different running times. Note that we *do not* compare the same cases between LU and INV – but given the small deviation between different batches – we feel that this still gives a representative picture.

The conclusion we draw from the measurement of the running time

	LU	INV
Batch 1	12.4901	5.6572
Batch 2	12.5437	5.7490
Batch 3	12.4545	5.7316

Table 6.4.: Gives the total running time in seconds for batches of 50000 cases each. The entries are chosen in the interval $[-10, 10]$. We compare the LU version to the INV version.

is that the INV version is more than twice a fast as the LU version. This is no reason for alarm, however, because this is expected, given the setup. The cases are small, and the overhead of having to do the bookkeeping of indices in the sparse case has a large impact in this setting. The running time of an individual case is small; in the order of a few dozen microseconds.

If we enlarge the range from which to choose the random entries to $[-1000, 1000]$, we can see the bookkeeping penalty decrease as the computational effort shifts from iteration handling to actual computations involving numbers. See Table 6.5 for a breakdown.

	LU	INV
Batch 1	37.2128	25.1496
Batch 2	37.3286	25.5482
Batch 3	37.1586	25.0753

Table 6.5.: Gives the total running time in seconds for batches of 50000 cases each. The entries are chosen in the interval $[-1000, 1000]$. We compare the LU version with the INV version.

If we go one step further and also increase the number of variables and constraints of the instances, as we will do in the following section, we can see that the situation shifts in favor of the LU version.

6.2.2. Medium Cases

We keep the general setup of the last section. That is, all entries and relations are chosen randomly. The entries are chosen from the interval $[-10, 10]$ in all cases presented here. This time, however, we vary the size of the problems generated. The number of variables n as well as the number of constraints m are chosen uniformly at random from an interval $[1, s]$, for each test case individually, where s is an integer. We present results for $s = 10, 30, 60$.

First, let us note that the structure of the solution landscape becomes more diverse. We start to see differences between the four different solution methods. This is summarized in Figure 6.6.

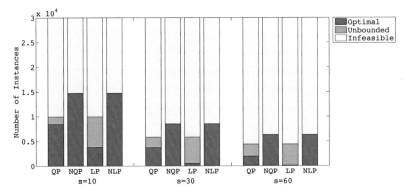

Figure 6.6.: Shows the fractions of *optimal*, *infeasible*, and *unbounded* problems. The groups of bars pertain to the different values of s. Within the group we indicate the four different solution methods; QP for quadratic program and LP for linear program. The N in NQP and NLP indicates nonnegativity bounds on the variables. For each value of s we consider 30000 cases.

We can see that, in general, the number of feasible solutions is decreasing with a growing number of constraints and variables. The following table summarizes the results about the running time.

	s=10		s=30		s=60	
	LU	INV	LU	INV	LU	INV
Batch 1	10.542	8.170	45.145	56.644	136.09	267.28
Batch 2	10.476	8.332	45.013	57.951	133.57	274.86
Batch 3	10.460	8.326	44.803	57.110	134.57	271.62

Table 6.7.: Gives the total combined running times in seconds for batches of 5000 cases each. For different values of s different instances are generated. The entries are chosen in the interval $[-10, 10]$. We compare the LU version with the INV version.

The relative advantage of the LU version becomes larger as s grows. This has two reasons. First, the bookkeeping of iteration overhead becomes less pronounced as the problem size grows. We believe this is due to the decreasing ratio between the expense of creating iterators and actually iterating over the entries. Second, the growing number of infeasible problems plays a role too. For these problems, the simplex algorithm never gets beyond its first phase, in which a linear program has to be solved. Because of the special structure of our factorization procedure we store the matrix D fully, even though it is symmetric. This is not the case for the INV variant. It seems favorable for the LU variant to finish in phase one because only the matrix A will have to be considered.

6.3. Polytope Distance

The mathematical description and comments about the implementation can be found in Section 5.4. Briefly speaking, we have to determine the minimum distance between the convex hulls of two point sets P and Q. This is implemented as part of the *Optimal Distances*[50] package in CGAL. That implementation fully relies on the quadratic programming solver. It is therefore an ideal test candidate. Evidently, when redesigning the quadratic programming solver, it should be an important goal that dependent applications do not suffer (too much) from this redesign.

The first test series we will discuss is depicted in Figure 6.8. We have used a homogeneous geometry kernel in CGAL to randomly create instances of the polytope distance problem with varying dimension. We measured the running times of the LU variant compared to the INV variant. We ran the same test setup for a Cartesian kernel too. Even though the running times were different (on average a little slower for Cartesian), looking at the comparison between LU and INV does not reveal any major differences. Therefore, we refrain from including the Cartesian results.

The two point sets are created as follows: Coordinates of the points in P are sampled uniformly at random from the interval $[0, 1048576]$ ($1048576_{\mathrm{dec}} \hat{=} 100000_{\mathrm{hex}}$). The point set P lies in the positive orthant. Coordinates of the points in Q are sampled from the interval $[-1048576, 0]$. The point set Q lies in the negative orthant. The convex hulls of P and Q are disjoint, and the minimum distance between them is positive (except for the unlikely case that both point sets happen to include the origin).

The purpose of this test is to see whether the LU variant offers any benefits for high dimensions, and indeed it does. Even though the LU variant is slower for dimensions $d \lesssim 300$, for $d \gtrsim 300$ the advantage of the sparse formulation takes over. The reason for this clearly lies

---

[50] http://www.cgal.org/Manual/latest/doc_html/cgal_manual/Polytope_distance_d/Chapter_main.html

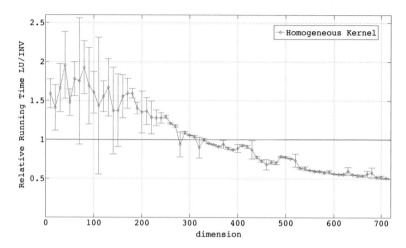

Figure 6.8.: Polytope distance problem with $|P| = |Q| = 1000$ in all test cases. The coordinate density is 1. The diamonds indicate average values of three test runs with a fixed dimension. The vertical bars indicate the standard deviation among runs with the same setup. The dimension was changed in increments of 10. The horizontal blue line represents the equality of the two variants; below LU is faster, above INV is faster.

in the sparse implementation of the matrix D; see Section 5.4. If we consider Theorem 3.5, we can deduce that the maximum basis size in this problem is $|B| \leq m + \text{rank}(D) = (d+2) + 2d = 3d + 2$. Assuming that increasing d (and keeping n fixed) leads to larger bases, we can conclude that matrix $D_{B,B}$ will have a growing number of zero rows and columns as d grows. The LU variant takes advantage of this fact and will skip all these entries, while the INV variant will always have to consider the zeros explicitly. A similar argument is true for the matrix A, or in other words yet, the density of the whole basis matrix decreases as the ratio d/n grows.

In Figure 6.9, we can see the actual running times corresponding to the relative running times in Figure 6.8. Even though the LU variant is slower for low dimensions, we notice that the absolute running times are not large. For $d \gtrsim 300$, where the LU variant becomes faster, the relative advantage seems more important because the running times become prohibitively high.

Note that the number of points in this test ($n = 1000$) is relatively low, and that the break-even point of the LU variant will be different for different values of n. We can see this in the next test series, for example.

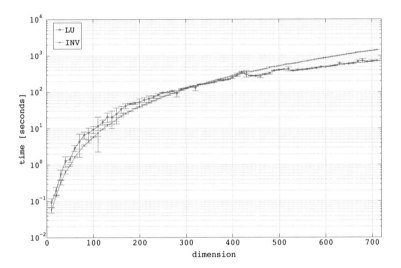

Figure 6.9.: Running times for the same cases as depicted Figure 6.8, that is, polytope distance problems with $|P| = |Q| = 1000$ and coordinate density 1.

Another setup we looked at is using a fixed dimension $d = 3$ but a variable number of points; see Figure 6.10. This is an important case, because real-world geometrical problems are formulated in physical space. Of course, it is desirable that the new version does not loose too

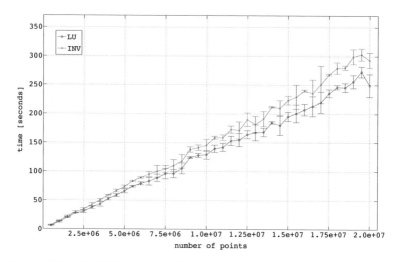

Figure 6.10.: Variable number of points polytope distance problem with $d = 3$ in all test cases.

much compared to the old one in this case. According to the previous test, this is not to be taken for granted, especially for low dimensions. The results that are depicted in Figure 6.10 are reassuring, however. We tested both variants with the number of points ranging between $5 \cdot 10^5$ and $2 \cdot 10^7$. The LU version is faster by more than 10% on average.

The final test is designed to show that sparse coordinates positively influence the LU variant. Recall that the matrix A (see equation (5.19)) containing the points is not implemented in a sparse fashion, because a priori we cannot have any sparsity information about the point sets P and Q, and doing online checking or preprocessing may be expensive. Nevertheless, the effect of thinning out the coordinates of the points is clearly giving the LU implementation an advantage, as we can see in Figure 6.11. In this test, saying that the coordinate density is $0 \leq p \leq 1$ means that a particular entry has probability p of being nonzero. The test is carried out for $d = 300$ and $n = 1000$, i.e., the same parameters as used in the first test.

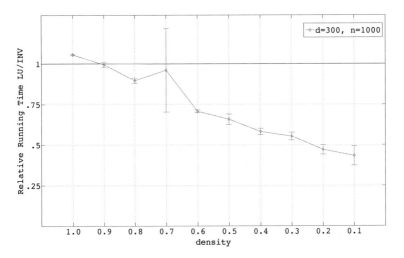

Figure 6.11.: Variable coordinate density polytope distance problem with $|P| = |Q| = 1000$ in all test cases.

The reason for this behavior is that the basis matrix is set up in a way to discard zero entries even if the input interface does not respect that. That is, during the computation we extract entries from the input iterators. Whenever we encounter a zero entry, we will *not* incorporate this entry into the basis matrix, which is stored internally.

6.4. Minimum Annulus

In this section we will review some results about the minimum annulus problem. We have already introduced the problem definition in

Section 5.5. Recall that the minimum annulus problem is formulated as a linear program, and that the constraint matrix is not sparse in general, because A mainly consists of the coordinates of the input points (see equation (5.28)). In these test sets we have used both a Cartesian as well as a homogeneous geometry kernel. Let us get to the results.

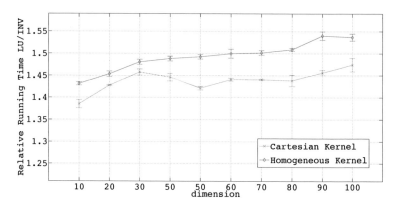

Figure 6.12.: Three minimum annulus problems are solved for each data point with variable dimension d, between 10 and 100. The number of points is $n = 1000$.

As we have described in Section 5.5 the implementation of the minimum annulus problem itself remains unchanged, and the adaption to the new LU version of the solver is done internally by wrapping a sparse iterator around the input iterator provided by the minimum annulus formulation. This comes at an expected penalty, as we can see in Figures 6.12 and 6.13.

In the former test we fix the number of points, $n = 1000$, and increase the dimension of the points. In the latter we fix the dimension $d = 3$ to and vary the number of points.

The good news is that in the case of $d = 3$ the penalty is not restrictive, and that in both tests the variability with respect to the main parameter is more or less stable. It would be interesting to try a sparse input scheme for the minimum annulus problem – as in the case of polytope distance – to see whether the running time improves.

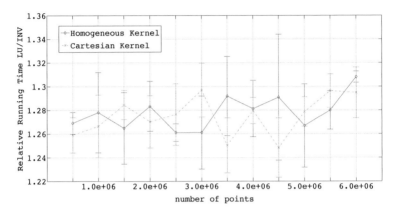

Figure 6.13.: Three minimum annulus problems are solved for each data point with a variable number of points, between $5 \cdot 10^5$ and $6 \cdot 10^6$. The geometric dimension is $d = 3$.

6.5. Extreme Points

In his master's thesis [71] Helbling discusses an application of linear programming that we are also going to have a closer look at. In particular, we will investigate a few cases, and try to see how the new LU version of the solver compares to the old INV version. The application in question is finding the set of extreme points of a point set in high dimensions.

Problem 6.14. *Given a set of points $P \in \mathbb{R}^d$, determine the subset $E \subseteq P$ of extreme points.*

To describe what an extreme point is, we need the concept of a convex hull of a set of points, which we defined earlier; see Definition 5.16.

Definition 6.15. *A point p in P is said to be* extreme *if it is not contained in the convex hull of the other points, i.e.,*

$$p \in P \text{ is extreme } \Leftrightarrow p \notin \text{conv}(P \setminus \{p\}).$$

There is an output-sensitive algorithm by Dulá and Helgason [45] which computes the extreme points without having to compute the convex hull of a point set. This is essential, because computing the convex hull is potentially expensive in higher dimensions. With respect to the dimension d, the convex hull may contain exponentially many facets. Therefore, any algorithm based on the computation of the convex hull inherently suffers from the danger of having an exponential running time. Roughly speaking, the Dulá-Helgason algorithm avoids computing the convex hull by performing a number of *convex combination tests*. The convex combination test is a procedure which checks whether a point p is the convex combination of some set of points $\mathcal{A} = \{a_1, \ldots, a_n\}$. This can be formulated as the following linear program:

$$
\begin{aligned}
\text{(EP)} \quad \min \quad & 0 \\
\text{s.t.} \quad & \sum_{i=1}^{n} a_i x_i = p \\
& \sum_{i=1}^{n} x_i = 1 \\
& x_i \geq 0, \qquad \forall 1 \leq i \leq n
\end{aligned}
\tag{6.16}
$$

It is a pure feasibility problem because the objective function is constant. Note that the columns of the constraint matrix

$$A = \left(\begin{array}{ccc} a_1 & \cdots & a_n \\ \hline 1 & \cdots & 1 \end{array} \right)$$

consist of the points in $\mathcal{A}$, and that the last row of A is a row of ones. The matrix A has $d+1$ rows and n columns. Using this, the linear programming formulation simplifies to the standard form,

$$
\begin{aligned}
\text{(EP)} \quad & \min && 0 \\
& \text{s.t.} && Ax = \left(\begin{array}{c} p \\ 1 \end{array} \right) && \text{(6.17)} \\
& && x_i \geq 0, && \forall 1 \leq i \leq n,
\end{aligned}
$$

where $x = (x_1, \ldots, x_n)^T$.

It should be clear that – if we choose points with few nonzero coordinates – we end up with a sparse matrix A. In the following we will look at the particular model where each of the points has exactly $k \geq 2$ nonzeros. This is also considered by Helbling [71]. We are not going to go deeper into the workings of the Dulá-Helgason algorithm though. The interested reader is referred to Helbling's thesis.

Helbling has implemented the mentioned algorithm as a package of CGAL that directly relies on the quadratic programming solver. So, it is easy and convenient for us to compare the old version of the solver (INV) with the new version (LU), which is optimized for sparse inputs. The first example we look at is computing the extreme points of one thousand points ($n = 1000$) with exactly two nonzeros each ($n_{\mathrm{nz}} = 2$). The value of the geometric dimension varies; ranging from 10 to 150 in increments of 5.

For each value of d we generated 3 random instances. The position of the nonzeros is chosen uniformly at random among all pairs of positions $\left(\binom{[n]}{2}\right)$ within each point, and the actual entries are chosen uniformly at random between -65535 and 65535 (which corresponds to the hexadecimal number $FFFF_{\text{hex}}$, or decimal $2^{16} - 1$). Figures 6.18 and 6.19 show the comparison of the running times.

There are two things that we notice. First, the running time of the LU variant stays virtually constant for increasing dimension while the INV variant increases; in particular for $d \gtrsim 70$. Therefore, the relative running time comparison in Figure 6.19 decreases further and further towards the right.

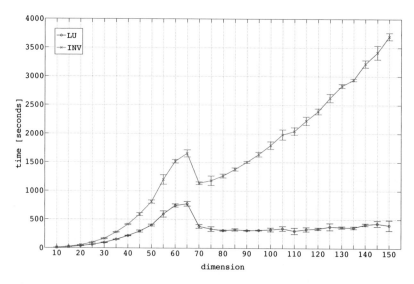

Figure 6.18.: Running 3 random cases with 1000 points each, and different values for $10 \leq d \leq 150$. The points have exactly 2 nonzero entries.

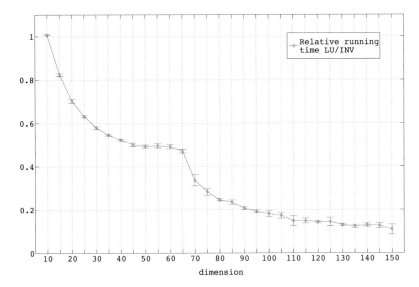

Figure 6.19.: This figure depicts the relative running times for the same cases displayed in Figure 6.18.

The second thing that we notice is the bump and sharp drop in running time that happens around $d \approx 65$. This can be – at least partially – explained by the following figure, Figure 6.20, which plots the number of extreme points found in the random test cases. It becomes clear that for $d \gtrsim 70$ almost all the points are extreme.

Intuitively, this makes sense. Consider some point having an extremal value in any of the coordinates. Then it is an extreme point, because it cannot be expressed as a convex combination of the remaining points. The higher the dimension, the higher the probability that some point has an extremal coordinate in at least one place.

The two findings seem to be linked, and the experimental data suggests, that beyond $d \gtrsim 70$ we are able to reduce the performance ratio almost arbitrarily with increasing dimension.

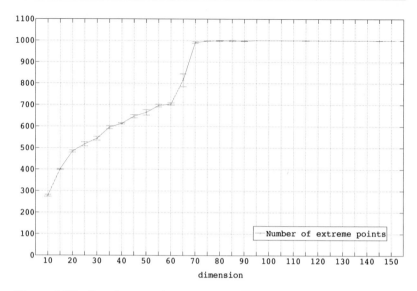

Figure 6.20.: For the same instances as in Figure 6.18 and Figure 6.19 we plot the number of extreme points. Recall that the total number of points is 1000 and that they have exactly 2 nonzero entries.

The Dulá-Helgason algorithm is output sensitive, i.e., its combinatorial running time depends on the final number of extreme points. This explains why the LU variant remains at a nearly fixed running time in the setup we tested, because both the number of convex combination tests *and* the number of nonzero entries in each of these remain constant. The INV variant suffers from the increasing overall dimension of A, in spite of most of the entries being zero.

Appendix

A.1. Encoding Numbers

Here are some basic facts about the encoding length of numbers. For proofs and more detail see [67]. We assume that integer numbers are encoded in binary encoding as $\{0, 1\}$ bit strings. By $\langle n \rangle$ we denote the encoding length of some integer.

The encoding length of an integer $n \in \mathbb{Z} \backslash \{0\}$ is

$$\langle n \rangle := 1 + \lceil \log_2(|n| + 1) \rceil. \tag{A.1}$$

We have to add one count for storing the sign. Encoding a plain 0 does not need that; one bit is enough.

Any rational number $r \in \mathbb{Q}$ can be uniquely expressed in co-prime form as p/q with $q \in \mathbb{N} \backslash \{0\}$ and $p \in \mathbb{Z}$. The encoding length is therefore given as

$$\langle r \rangle := \langle p \rangle + \langle q \rangle . \tag{A.2}$$

Strictly speaking, we are over-counting on two accounts, but for the following considerations we can disregard this. On the one hand, as we already pointed out, we do not need to store the sign of p if it is 0. On the other hand, we never have to encode the sign for q, because it was defined as nonnegative.

The encoding length of a vector $\langle v \rangle$ or a matrix $\langle A \rangle$ is simply the sum of encoding lengths of its individual entries. The encoding length of a sum or a product (the same holds for difference and integral division) is nicely bounded. For $r, s \in \mathbb{Q}$ it holds that

$$\langle r + s \rangle \leq \langle r \rangle + \langle s \rangle , \tag{A.3}$$

$$\langle rs \rangle \leq \langle r \rangle + \langle s \rangle . \tag{A.4}$$

Finally, we point out an important fact about the encoding length of the determinant of a matrix.

Lemma A.5 (Lemma (1.3.4)(b) of [67]).
For every matrix $R \in \mathbb{Q}^{n \times n}$,

$$\langle \det(R) \rangle \leq 2 \langle R \rangle - n^2 . \tag{A.6}$$

The lemma shows that encoding the determinant is possible in space polynomial in the encoding length of the input matrix.

A.2. Models of Computation

The following definitions and conventions follow Section 1.3 of [67]. There are two different models of computation widely used. We assume that all numbers are encoded as binary strings, as described in Appendix A.1.

The *Turing machine model* counts the number of moves of the read-write head of the Turing machine. The complexity of an arithmetic operation like adding or multiplying two integer numbers is bounded by a polynomial of the encoding lengths of the two integers.

Sometimes it is more natural, however, to consider the *arithmetic model*, which counts the number of elementary arithmetic operations on real numbers such as addition, subtraction, multiplication, division, and comparison instead. In particular, if we are dealing with real-life computers where computations are often carried out with fixed precision (and the importance of the encoding length of the numbers involved takes the back seat) we only care about this type of operation count. In a broader context, this is also called the *unit cost model*, assuming that some elementary operation takes up one unit of time only.

The *input size* of an instance in the Turing machine model is the combined encoding length of all the numbers in the input. The input size of an instance in the arithmetic model disregards the encoding length of the actual numbers, and it only counts the number of input numbers.

There are problems that are polynomially bounded in the one model but not in the other one; and vice versa. For example, computing 2^{2^n} by repeated squaring is polynomial in the arithmetic model but not in the Turing machine model (because the output number has exponentially many digits). On the other hand, the well-known Euclidean algorithm to compute the greatest common divisor of two numbers is polynomial in the Turing machine model but not in the arithmetic model (because the input always consists of two numbers only).

If we say that an algorithm runs in polynomial time, we therefore have to specify what model we are referring to. We can refine this idea by supplementing the count of arithmetic operations by the assertion that the numbers involved in the computations do not grow too large, i.e., their encoding length is polynomially bounded by the encoding length of the input. Of course, if this assertion holds, we can execute an algorithm that is polynomial in the arithmetic model in polynomial time on a Turing machine too. We say that an algorithm runs in *strongly polynomial* time if it uses a polynomial number of elementary operations (in the arithmetic model) *and* if it uses polynomial space (in the Turing machine model). If an algorithm is polynomial in the arithmetic model but its running time depends on the encoding size of the input, we say that it runs in *weakly polynomial* time.

A.3. Geometry Basics

Let us gather a few basic definitions and tools about geometry, so we can refer to them in the main chapters of the present thesis.

In CGAL there are two basic types of geometry kernels available. The *Cartesian* and the *homogeneous* kernel. They differ in the way coordinates are represented, and in the requirements on the number types that have to be provided. In order to understand the difference better, let us explain how points are represented.

Arguably the most natural way to represent a point in $\mathbb{R}^d$ is using Cartesian coordinates. A (Cartesian) point p_C is uniquely defined by

d real numbers, $c_1, \ldots, c_d$, that specify its position with respect to the coordinate axes,

$$p_C := (c_1, \ldots, c_d).$$

Another possible way of representing points in $\mathbb{R}^d$ are homogeneous coordinates. In this system, a (homogeneous) point p_H is defined by $d + 1$ real numbers,

$$p_H := (h_1, \ldots, h_d, h).$$

The last coordinate h takes a special role, hence there is no subscript. Any given Cartesian point p_C is identified with $(h_1/h, \ldots, h_d/h)$, so that p_C corresponds to the tuple $(c_1 h, \ldots, c_d h, h)$ in homogeneous coordinates, for some $h \neq 0$. It should be clear that a point does not uniquely define its homogeneous coordinates, because multiplying all its (homogeneous) coordinates by some nonzero constant does not change the point represented.

The main motivation for homogeneous coordinates comes from projective geometry and from the need to be able to represent points at infinity, which is achieved by setting $h = 0$. There is another important advantage to the homogeneous system that can be of advantage in geometric computations. Because of the fact that h can be used as a normalizing constant, divisions can be avoided by multipling the last coordinate by the divisor. Transformation to Cartesian coordinates is achieved by dividing all the other coordinates by h.

Divisions are avoided in homogeneous geometry, which helps to save expensive operations. It also lets us loosen the requirements on the number type in a concrete implementation such as the geometric kernels of CGAL. A homogeneous kernel can operate using a ring number type, that *does not* provide a division operation.

A.4. Implementation of Doolittle's Algorithm

Algorithm 7: Doolittle's LU factorization

Input: $A \in \mathrm{M}_n(F)$

Output: $L \in \mathrm{L}_n(F)$, $U \in \mathrm{U}_n(F)$, s.t. $LU = A$.

1   **for** $i = 1$ *to* n **do**

2   **for** $j = 1$ *to* $i - 1$ **do**

3       $\alpha \leftarrow a_{i,j}$;

4   **for** $k = 1$ *to* $j - 1$ **do**

5         $\alpha \leftarrow \alpha - a_{i,k}a_{k,j}$;

6   **end**

7       $a_{i,j} \leftarrow \frac{\alpha}{a_{j,j}}$;

8   **end**

9   **for** $j = i$ *to* n **do**

10       $\alpha \leftarrow a_{i,j}$;

11   **for** $k = 1$ *to* $i - 1$ **do**

12         $\alpha \leftarrow \alpha - a_{i,k}a_{k,j}$;

13   **end**

14       $a_{i,j} \leftarrow \alpha$;

15   **end**

16   **end**

17   $L \leftarrow I_n + \mathrm{tril}(A, -1)$;

18   $U \leftarrow \mathrm{triu}(A)$;

A.5. Implementation of Crout's Algorithm

Algorithm 8: Crout's LU factorization

Input: $A \in \mathrm{M}_n(F)$

Output: $L \in \mathrm{L}_n(F)$, $U \in \mathrm{U}_n(F)$, s.t. $LU = A$.

1 **for** $j = 1$ **to** n **do**
2   **for** $i = j$ **to** n **do**
3         $\alpha \leftarrow a_{i,j}$;
4   **for** $k = 1$ **to** $j - 1$ **do**
5            $\alpha \leftarrow \alpha - a_{i,k} a_{k,j}$;
6   **end**
7         $a_{i,j} \leftarrow \alpha$;
8   **end**
9   **for** $j = j$ **to** n **do**
10        $\alpha \leftarrow a_{i,j}$;
11   **for** $k = 1$ **to** $i - 1$ **do**
12          $\alpha \leftarrow \alpha - a_{i,k} a_{k,j}$;
13   **end**
14        $a_{i,j} \leftarrow \alpha$;
15   **end**
16 **end**
17 $L \leftarrow I_n + \mathrm{tril}(A, -1)$;
18 $U \leftarrow \mathrm{triu}(A)$;

A.6. CGAL Documentation

Three pages, outlining the concept `QuadraticProgram`:

Concept

QuadraticProgram

Definition

A model of QuadraticProgram describes a convex quadratic program of the form

$$\text{(QP)} \quad \text{minimize} \quad \mathbf{x}^T D \mathbf{x} + \mathbf{c}^T \mathbf{x} + c_0$$
$$\text{subject to} \quad A\mathbf{x} \gtreqless \mathbf{b},$$
$$\mathbf{l} \le \mathbf{x} \le \mathbf{u}$$

in n real variables $\mathbf{x} = (x_0, \dots, x_{n-1})$. Here,

- A is an $m \times n$ matrix (the constraint matrix),
- $\mathbf{b}$ is an m-dimensional vector (the right-hand side),
- $\gtreqless$ is an m-dimensional vector of relations from $\{\le, =, \ge\}$,
- $\mathbf{l}$ is an n-dimensional vector of lower bounds for $\mathbf{x}$, where $l_j \in \mathbb{R} \cup \{-\infty\}$ for all j
- $\mathbf{u}$ is an n-dimensional vector of upper bounds for $\mathbf{x}$, where $u_j \in \mathbb{R} \cup \{\infty\}$ for all j
- D is a symmetric positive-semidefinite $n \times n$ matrix (the quadratic objective function),
- $\mathbf{c}$ is an n-dimensional vector (the linear objective function), and
- c_0 is a constant.

The description is given by appropriate *random-access* iterators over the program data, see below. The program therefore comes in *dense* representation which includes zero entries.

Has Models

Quadratic_program<NT>
Quadratic_program_from_mps<NT>
Quadratic_program_from_sparse_iterators<A_s_it, B_it, R_it, FL_it, L_it, FU_it, U_it, D_s_it, C_it>
Quadratic_program_from_iterators<A_it, B_it, R_it, FL_it, L_it, FU_it, U_it, D_it, C_it>

Types

QuadraticProgram:: A_sparse_iterator
    A random access iterator type to go columnwise over the constraint matrix A. The value type is an object that provides bidirectional sparse iterators for the column in question by member calls to *begin()* and *end()*. Such a column iterator *it* is sparse, providing *(index,value)* pairs of all non-zero elements of the column. The index is accessed by *it->first* and the value is accessed by *it->second*.

QuadraticProgram:: A_iterator
    A random access iterator type to go columnwise over the constraint matrix A. The value type is an random access iterator type for an individual column that goes over the entries in that column.

28

QuadraticProgram:: B_iterator		A random access iterator type to go over the entries of the right-hand side **b**.
QuadraticProgram:: R_iterator		A random access iterator type to go over the relations $\gtreqless$. The value type of *R_iterator* is *CGAL::Comparison_result*.
QuadraticProgram:: FL_iterator		A random access iterator type to go over the existence (finiteness) of the lower bounds $l_j, j = 0, \ldots, n-1$. The value type of *FL_iterator* is *bool*.
QuadraticProgram:: L_iterator		A random acess iterator type to go over the entries of the lower bound vector **l**.
QuadraticProgram:: UL_iterator		A random access iterator type to go over the existence (finiteness) of the upper bounds $u_j, j = 0, \ldots, n-1$. The value type of *UL_iterator* is *bool*.
QuadraticProgram:: U_iterator		A random acess iterator type to go over the entries of the upper bound vector **u**.
QuadraticProgram:: D_sparse_iterator		A random access iterator type to go rowwise over the matrix 2*D*. The value type is an object that provides bidirectional sparse iterators for the row in question by member calls to *begin()* and *end()*. Such a column iterator *it* is sparse, providing *(index,value)* pairs of all non-zero elements of the row. The index is accessed by *it->first* and the value is accessed by *it->second*.
QuadraticProgram:: D_iterator		A random access iterator type to go rowwise over the matrix 2*D*. The value type is a random access iterator type for an individual row that goes over the entries in that row, up to (and including) the entry on the main diagonal.
QuadraticProgram:: C_iterator		A random access iterator type to go over the entries of the linear objective function vector **c**.

Operations

int	*qp.get_n() const*	returns the number *n* of variables (number of columns of *A*) in *qp*.
int	*qp.get_m() const*	returns the number *m* of constraints (number of rows of *A*) in *qp*.
A_sparse_iterator	*qp.get_a_sparse() const*	
		returns an iterator over the columns of *A*. The corresponding past-the-end iterator is *get_a_sparse()+get_n()*. For $j = 0, \ldots, n-1$, $*(get\_a() + j)$ is an object that provides bidirectional sparse iterators for column *j* by calls to *begin()* and *end()*. These column iterators provide *(index,value)* pairs for all non-zero entries.

29

B_iterator	qp.get_b() const	returns an iterator over the entries of **b**. The corresponding past-the-end iterator is get_b()+get_m().
R_iterator	qp.get_r() const	returns an iterator over the entries of $\gtreqless$. The corresponding past-the-end iterator is get_r()+get_m(). The value CGAL::SMALLER stands for $\leq$, CGAL::EQUAL stands for $=$, and CGAL::LARGER stands for $\geq$.
FL_iterator	qp.get_fl() const	returns an iterator over the existence of the lower bounds $l_j, j = 0, \ldots, n-1$. The corresponding past-the-end iterator is get_fl()+get_n(). If *(get_fl()+j) has value true, the variable x_j has a lower bound given by *(get_l()+j), otherwise it has no lower bound.
L_iterator	qp.get_l() const	returns an iterator over the entries of **l**. The corresponding past-the-end iterator is get_l()+get_n(). If *(get_fl()+j) has value false, the value *(get_l()+j) is not accessed. *Precondition*: if both *(get_fl()+j) and *(get_fu()+j) have value true, then $*(get\_l()+j) \leq *(get\_u()+j)$
FU_iterator	qp.get_fu() const	returns an iterator over the existence of the upper bounds $u_j, j = 0, \ldots, n-1$. The corresponding past-the-end iterator is get_fu()+get_n(). If *(get_fu()+j) has value true, the variable x_j has an upper bound given by *(get_u()+j), otherwise it has no upper bound.
L_iterator	qp.get_u() const	returns an iterator over the entries of **u**. The corresponding past-the-end iterator is get_u()+get_n(). If *(get_fu()+j) has value false, the value *(get_u()+j) is not accessed. *Precondition*: if both *(get_fl()+j) and *(get_fu()+j) have value true, then $*(get\_l()+j) \leq *(get\_u()+j)$
D_sparse_iterator	qp.get_d_sparse() const	returns an iterator over the rows of $2D$. The corresponding past-the-end iterator is get_d_sparse()+get_n(). For $i = 0, \ldots, n-1$, $*(get\_d()+i)$ is an object that provides bidirectional sparse iterators for row i by calls to *begin()* and *end()*. These column iterators provide (*index,value*) pairs for all non-zero entries.
C_iterator	qp.get_c() const	returns an iterator over the entries of **c**. The corresponding past-the-end iterator is get_c()+get_n().
std::iterator_traits<C_iterator>::value_type		
	qp.get_c0() const	returns the constant term c_0 of the objective function.

Requirements

The value types of all iterator types (nested iterator types for A_iterator and D_iterator, and the types of *value* in the nested sparse iterators of A_sparse_iterator and D_sparse_iterator, respectively) must be convertible to some common *IntegralDomain ET*.

30

Bibliography

[1] ISO/IEC 14882:2011: Information Technology – Programming Languages – C++, 2011.

[2] ISO/IEC/IEEE 60559:2011 (IEEE Std 754-2008): Information Technology – Microprocessor Systems – Floating-Point arithmetic, 2011.

[3] Patrick R. Amestoy, Timothy A. Davis, and Iain S. Duff. An approximate minimum degree ordering algorithm. *SIAM Journal on Matrix Analysis and Applications*, 17(4):886–905, 1996.

[4] Patrick R. Amestoy, Timothy A. Davis, and Iain S. Duff. Algorithm 837: AMD, an approximate minimum degree ordering algorithm. *ACM Transactions on Mathematical Software*, 30(3):381–388, 2004.

[5] Patrick R. Amestoy and Chiara Puglisi. An unsymmetrized multifrontal LU factorization. *SIAM Journal on Matrix Analysis and Applications*, 24(2):553–569, 2002.

[6] Erling D. Andersen and Knud D. Andersen. Presolving in linear programming. *Mathematical Programming*, 71(2):221–245, 1995.

[7] Erling D. Andersen and Knud D. Andersen. The MOSEK interior point optimizer for linear programming: an implementation of the homogeneous algorithm. In H. Frenk *et al.*, editor, *High Performance Optimization*, pages 197–232. Kluwer Academic Publishers, Dordrecht, Netherlands, 2000.

[8] Erling D. Andersen, Cees Roos, and Tamas Terlaky. On implementating a primal-dual interior-point method for conic quadratic optimization. *Mathematical Programming (Series B)*, 95(2):249–277, 2003.

[9] Rainer Bacher and Hans P. Van Meeteren. Security dispatch based on coupling of linear and quadratic programming techniques. In A. J. Calvaer, editor, *Power Systems: Modelling and Control Applications. Proceedings of the IFAC Symposium*, pages 211–217. Pergamon Press, Oxford, England, 1989.

[10] David H. Bailey, King Lee, and Horst D. Simon. Using Strassen's algorithm to accelerate the solution of linear systems. *The Journal of Supercomputing*, 4(4):357–371, 1991.

[11] Erwin H. Bareiss. Sylvester's identity and multistep preserving Gaussian elimination. *Mathematics of Computation*, 22(103):565–578, 1968.

[12] Erwin H. Bareiss. Computational solution of matrix problems over an integral domain. *Journal of the Institute of Mathematics and its Applications*, 10:68–104, 1972.

[13] Richard H. Bartels and Gene H. Golub. The simplex method of linear programming using LU decomposition. *Communications of the ACM*, 12(5):266–268, 1969.

[14] Maurice S. Bartlett. An inverse matrix adjustment arising in discriminant analysis. *The Annals of Mathematical Statistics*, 22(1):107–111, 1951.

[15] Gerard Bashein and Mark Enns. Computation of optimal controls by a method combining quasi-linearization and quadratic programming. *International Journal of Control*, 16(1):177–187, 1972.

[16] Laleh Behjat and Anthony Vannelli. VLSI concentric partitioning using interior point quadratic programming. In *Proceedings of the IEEE International Symposium on Circuits and Systems (ISCAS), Piscataway, NJ, USA*, volume 6, pages 93–96, 1999.

[17] Alberto Bemporad, Manfred Morari, Vivek Dua, and Efstratios N. Pistikopoulos. The explicit solution of model predictive control via multiparametric quadratic programming. In *Proceedings of the American Control Conference (ACC), Danvers, MA, USA*, volume 2, pages 872–876, 2000.

[18] John M. Bennet. Triangular factors of modified matrices. *Numerische Mathematik*, 7(3):216–221, 1965.

[19] O. Bertoldi, M. V. Cazzol, A. Garzillo, and M. Innorta. A dual quadratic programming algorithm oriented to the probabilistic analysis of large interconnected networks. In *Proceedings of the 12th Power Systems Computation Conference (PSCC), Zürich, Switzerland*, volume 2, pages 1249–1255, 1996.

[20] Michael J. Best and Jivendra Kale. Quadratic programming for large-scale portfolio optimization. In J. Keyes, editor, *Financial Services Information Systems*, pages 513–529. CRC Press, Boca Raton, FL, USA, 2000.

[21] Stephen Boyd and Lieven Vandenberghe. *Convex Optimization*. Cambridge University Press, Cambridge, England, 2004.

[22] Yves Brise and Bernd Gärtner. Clarkson's algorithm for violator spaces. *Computational Geometry*, 44(2):70–81, 2011.

[23] James R. Bunch and John E. Hopcroft. Triangular factorization and inversion by fast matrix multiplication. *Mathematics of Computation*, 28(125):231–236, 1974.

[24] J. L. Carpentier, G. Cotto, and P. L. Niederlander. New concepts for automatic generation control in electric power systems using parametric quadratic programming. In A. Alonso-Concheiro, editor, *Real Time Digital Control Applications. Proceedings of the IFAC/IFIP Symposium*, pages 595–600. Pergamon Press, Oxford, England, 1984.

[25] Jordi Castro. Quadratic interior-point methods in statistical disclosure control. *Computational Management Science*, 2(2):107–121, 2005.

[26] Václav Chvátal. *Linear Programming*. W. H. Freeman and Company, New York, NY, USA, 1983.

[27] Kenneth L. Clarkson. Las Vegas algorithms for linear and integer programming when the dimension is small. *Journal of the ACM*, 42(2):488–499, 1995.

[28] G. C. Contaxis, C. Delkis, E. N. Glytsis, and B. C. Papadias. Economic power dispatch with line security limits using Z-matrix techniques and quadratic programming. In E. Lauger and J. Moltoft, editors, *Reliability in Electrical and Electronic Components and Systems. Proceedings of the 5th European Conference on Electrotechnics (EUROCON), Copenhagen, Denmark*, pages 709–713, 1982.

[29] Samuel D. Conte and Carl de Boor. *Elementary Numerical Analysis*. McGraw-Hill, New York, NY, USA, 1980.

[30] Don Coppersmith and Shmuel Winograd. Matrix multiplication via arithmetic progressions. *Journal of Symbolic Computation*, 9(3):251–280, 1990.

[31] Robert M. Corless and David J. Jeffrey. The Turing factorization of a rectangular matrix. *SIGSAM Bulletin (ACM Special Interest Group on Symbolic and Algebraic Manipulation)*, 31(3):20–28, 1997.

[32] J. S. Cotner and Reuven R. Levary. A quadratic programming model for determining short-term multiple currency portfolios. *Opsearch*, 24(4):218–227, 1987.

[33] George B. Dantzig. *Linear programming and extensions*. Princeton University Press, Princeton, NJ, USA, 1963.

[34] George B. Dantzig and William Orchard-Hays. The product form for the inverse in the simplex method. *Mathematical Tables and Other Aids to Computation*, 8(46):64–67, 1954.

[35] Timothy A. Davis. Algorithm 832: UMFPACK V4.3 - an unsymmetric-pattern multifrontal method. *ACM Transactions on Mathematical Software*, 30(2):196–199, 2004.

[36] Timothy A. Davis. A column pre-ordering strategy for the unsymmetric-pattern multifrontal method. *ACM Transactions on Mathematical Software*, 30(2):165–195, 2004.

[37] Timothy A. Davis. *Direct methods for sparse linear systems*. Fundamentals of algorithms. Society for Industrial and Applied Mathematics (SIAM), Philadelphia, PA, USA, 2006.

[38] Timothy A. Davis and Iain S. Duff. An unsymmetric-pattern multifrontal method for sparse *LU* factorization. *SIAM Journal on Matrix Analysis and Applications*, 18(1):140–158, 1997.

[39] Timothy A. Davis and Iain S. Duff. Combined unifrontal/multifrontal method for unsymmetric sparse matrices. *ACM Transactions on Mathematical Software*, 25(1):1–20, 1999.

[40] Timothy A. Davis, John R. Gilbert, Stefan I. Larimore, and Esmond G. Ng. Algorithm 836: COLAMD, a column approximate minimum degree ordering algorithm. *ACM Transactions on Mathematical Software*, 30(3):377–380, 2004.

[41] Timothy A. Davis, John R. Gilbert, Stefan I. Larimore, and Esmond G. Ng. A column approximate minimum degree ordering algorithm. *ACM Transactions on Mathematical Software*, 30(3):353–376, 2004.

[42] Iain S. Duff, Albert M. Erisman, and John K. Reid. On George's nested dissection method. *SIAM Journal on Numerical Analysis*, 13(5):686–695, 1976.

[43] Iain S. Duff, Albert M. Erisman, and John K. Reid. *Direct Methods for Sparse Matrices*. Clarendon Press, Oxford, England, 1986.

[44] Iain S. Duff and John K. Reid. The multifrontal solution of indefinite sparse symmetric linear systems. *ACM Transactions on Mathematical Software*, 9(3):302–325, 1983.

[45] José H. Dulá and Richard V. Helgason. A new procedure for identifying the frame of the convex hull of a finite collection of points in

multidimensional space. *European Journal of Operational Research*, 92(2):352–367, 1996.

[46] David Dureisseix. Generalized fraction-free *LU* factorization for singular systems with kernel extraction. *Linear Algebra and its Applications*, 436(1):27–40, 2012.

[47] Jack Edmonds. Systems of distinct representatives and linear algebra. *Journal of Research of the National Bureau of Standards*, 71B(4):241–245, 1967.

[48] Jack Edmonds and Jean-François Maurras. Note sur les *Q*-matrices d'Edmonds. *Recherche Opérationnelle*, 31(2):203–209, 1997.

[49] Andreas Fabri, Geert-Jan Giezeman, Lutz Kettner, Stefan Schirra, and Sven Schönherr. On the design of CGAL a computational geometry algorithms library. *Software: Practice and Experience*, 30(11):1167–1202, 2000.

[50] Ji-Yuan Fan and Lan Zhang. Real-time economic dispatch with line flow and emission constraints using quadratic programming. *IEEE Transactions on Power Systems*, 13(2):320–325, 1998.

[51] Hassan A. Farhat and Steven G. From. A quadratic programming approach to estimating the testability and random or deterministic coverage of a VLSI circuit. *VLSI Design*, 2(3):223–231, 1994.

[52] Hassan A. Farhat, Steven G. From, and Antonio Lioy. A quadratic programming approach to estimating the testability and coverage distributions of a VLSI circuit. *Microprocessing and Microprogramming*, 35(1–5):479–483, 1992.

[53] Kaspar Fischer, Bernd Gärtner, and Martin Kutz. Fast smallest-enclosing-ball computation in high dimensions. In G. Di Battista and U. Zwick, editors, *Proceedings of the 11th Annual European Symposium on Algorithms (ESA), Budapest, Hungary*, volume 2832 of *Lecture Notes in Computer Science*, pages 630–641. Springer, Berlin, Germany, 2003.

[54] John J. H. Forrest and John A. Tomlin. Updated triangular factors of the basis to maintain sparsity in the product form simplex method. *Mathematical Programming*, 2(1):263–278, 1972.

[55] Michael Garey and David S. Johnson. *Computers and Intractability: A Guide to the Theory of NP-completeness.* W.H. Freeman and Company, New York, NY, USA, 1979.

[56] Bernd Gärtner. Exact arithmetic at low cost – a case study in linear programming. *Computational Geometry*, 13(2):121–139, 1999.

[57] Bernd Gärtner, Jiří Matoušek, Leo Rüst, and Petr Škovroň. Violator spaces: structure and algorithms. *Discrete Applied Mathematics*, 156(11):2124–2141, 2008.

[58] Bernd Gärtner and Sven Schönherr. An efficient, exact, and generic quadratic programming solver for geometric optimization. In *Proceedings of the Symposium on Computational Geometry (SoCG)*, pages 110–118, 2000.

[59] Bernd Gärtner and Emo Welzl. Linear programming – randomization and abstract frameworks. In *Proceedings of the 13th Annual Symposium on Theoretical Aspects of Computer Science (STACS)*, volume 1046 of *Lecture Notes in Computer Science*, pages 669–687. Springer-Verlag, Berlin, Germany, 1996.

[60] Bernd Gärtner and Emo Welzl. A simple sampling lemma: analysis and applications in geometric optimization. *Discrete & Computational Geometry*, 25(4):569–590, 2001.

[61] Alan George. Nested dissection of a regular finite element mesh. *SIAM Journal on Numerical Analysis*, 10(2):345–363, 1973.

[62] Alan George. An automatic one-way dissection algorithm for irregular finite element problems. *SIAM Journal on Numerical Analysis*, 17(6):740–751, 1980.

[63] Philip E. Gill, Gene H. Golub, Walter Murray, and Michael A. Saunders. Methods for modifying matrix factorizations. *Mathematics of Computation*, 28(126):505–535, 1974.

[64] Philip E. Gill, Walter Murray, and Michael A. Saunders. SNOPT: An SQP algorithm for large-scale constrained optimization. *SIAM Journal on Optimization*, 12(4):979–1006, 2002. Updated version appeared in *SIAM Review*, 47(1):99–131, 2005.

[65] Gene H. Golub and Charles F. Van Loan. *Matrix Computations*. Johns Hopkins University Press, Baltimore, MD, USA, 3rd edition, 1996.

[66] Nicholas Gould and Philippe Toint. A quadratic programming page. http://www.numerical.rl.ac.uk/qp/qp.html.

[67] Martin Grötschel, László Lovász, and Alexander Schrijver. *Geometric Algorithms and Combinatorial Optimization*. Springer-Verlag, Berlin, Germany, 1988.

[68] Z. Guo and G. Xu. Calculation of economic dispatch of interconnected system using quadratic programming. *Automation of Electric Power Systems*, 22(1):40–44, 1998.

[69] Fred G. Gustavson, Werner Liniger, and Ralph A. Willoughby. Symbolic generation of an optimal Crout algorithm for sparse systems of linear equations. *Journal of the ACM*, 17(1):87–109, 1970.

[70] William W. Hager. Updating the inverse of a matrix. *SIAM Review*, 31(2):221–239, 1989.

[71] Christian Helbling. Extreme points in medium and high dimensions. Master's thesis, ETH Zürich, Switzerland, 2010.

[72] Guo H. Huang, Brian W. Baetz, and Gilles G. Patry. Waste flow allocation planning through a grey fuzzy quadratic-programming approach. *Civil Engineering Systems*, 11(3):209–243, 1994.

[73] M. Huang. Linear and quadratic programming methods for solving security-constrained economic dispatch problems. Master's thesis, Department of Electrical and Computer Engineering, University of Waterloo, Ontario, Canada, 1991.

[74] Hanh Huynh. *A Large-Scale Quadratic Programming Solver Based on Block-LU Updates of the KKT System.* PhD thesis, Stanford University, CA, USA, 2008.

[75] Bruce M. Irons. A frontal solution program for finite element analysis. *International Journal for Numerical Methods in Engineering*, 2:5–32, 1970.

[76] Dror Irony, Gil Shklarski, and Sivan Toledo. Parallel and fully recursive multifrontal sparse Cholesky. *Future Generation Computer Systems*, 20(3):425–440, 2004.

[77] Hussein Jaddu and Etsujiro Shimemura. Computation of optimal control trajectories using Chebyshev polynomials: parametrization, and quadratic programming. *Optimal Control Applications and Methods*, 20(1):21–42, 1999.

[78] Richard M. Karp. Reducibility among combinatorial problems. In R. E. Miller and J. W. Thatcher, editors, *Complexity of Computer Computations*, pages 85–103. Plenum Press, New York, NY, USA, 1972.

[79] Andrew A. Kennings and Anthony Vannelli. VLSI placement using quadratic programming and network partitioning techniques. *International Transactions in Operational Research*, 4(5–6):353–364, 1997.

[80] L. G. Khachiyan. A polynomial algorithm for linear programming. *Soviet Mathematics Doklady*, 20:191–194, 1979.

[81] Euntai Kim, Hyung-Jin Kang, and Mignon Park. Numerical stability analysis of fuzzy control systems via quadratic programming and linear matrix inequalities. *IEEE Transactions on Systems, Man and Cybernetics, Part A (Systems and Humans)*, 29(4):333–346, 1999.

[82] Victor Klee and George J. Minty. How good is the simplex algorithm? In *Proceedings of the 3rd Symposium on Inequalities, University of California, Los Angeles, CA, USA*, pages 159–175. Academic Press, Waltham, MA, USA, 1972.

[83] Jürgen M. Kleinhans, Georg Sigl, Frank M. Johannes, and Kurt J. Antreich. GORDIAN: VLSI placement by quadratic programming and slicing optimization. *IEEE Transactions on Computer Aided Design of Integrated Circuits and Systems*, 10(3):356–365, 1991.

[84] L. B. Kovacs and M. Kotel. Indefinite quadratic programming by gradient projection method and its application to optimal control of a chemical plant. In *Proceedings of the IFAC Symposium on Multivariable Control Systems VDI/VDE, Fachgruppe Reqelungstechnik, Düsseldorf, West Germany*, 1968.

[85] M. K. Kozlov, S. P. Tarasov, and L. G. Khachiyan. Polynomial solvability of convex quadratic programming. *USSR Computational Mathematics and Mathematical Physics*, 20(5):223–228, 1980.

[86] Piyush Kumar, Joseph S. B. Mitchell, and E. Alper Yıldırım. Computing core-sets and approximate smallest enclosing hyperspheres in high dimensions. In *Proceedings of the International Workshop on Algorithm Engineering and Experimentation (ALENEX), Baltimore, MD, USA*, pages 45–55, 2003.

[87] Yen-Tai Lai, Chi-Chou Kao, and Wu-Chien Shie. A quadratic programming method for interconnection crosstalk minimization. In *Proceedings of the IEEE International Symposium on Circuits and Systems (ISCAS), Piscataway, NJ, USA*, volume 6, pages 270–273, 1999.

[88] Hong R. Lee and David Saunders. Fraction-free Gaussian elimination for sparse matrices. *Journal of Symbolic Computation*, 19(5):393–402, 1995.

[89] Donald J. Leo and Daniel J. Inman. A quadratic programming approach to the design of active-passive vibration isolation systems. *Journal of Sound and Vibration*, 220(5):807–825, 1999.

[90] Bosco H. Leung. Design methodology of decimation filters for over-sampled ADC based on quadratic programming. *IEEE Transactions on Circuits and Systems*, 38(10):1121–1132, 1991.

[91] Chang-Tsuo Liu, Gabor C. Temes, and Henry Samueli. FIR filter design using quadratic programming. *IEEE International Symposium on Circuits and Systems (ISCAS), New York, NY, USA*, 1:148–151, 1991.

[92] Joseph W.-H. Liu. The multifrontal method for sparse matrix solution: theory and practice. *SIAM Review*, 34(1):82–109, 1992.

[93] Xiang Liu, Youxian Sun, and Wenhai Wang. Stabilizing control of robustness for systems with maximum uncertain parameters – a quadratic programming approach. *Control Theory and Applications*, 16(5):729–732, 1999.

[94] Christopher Maes. *A Regularized Active-Set Method for Sparse Convex Quadratic Programming*. PhD thesis, Stanford University, CA, USA, 2010.

[95] Kazimierz Malanowski. On application of a quadratic programming procedure to optimal control problems in systems described by parabolic equations. *Control and Cybernetics*, 1(1–2):43–56, 1972.

[96] Harry M. Markowitz. Portfolio selection. *Journal of Finance*, 7(1):77–91, 1952.

[97] Harry M. Markowitz. The optimization of a quadratic function subject to constraints. *Naval Research Logistics Quarterly*, 3(1–2):111–133, 1956.

[98] Harry M. Markowitz. The elimination form of the inverse and its application to linear programming. *Management Science*, 3(3):255–269, 1957.

[99] Jiří Matoušek. Removing degeneracy in LP-type problems revisited. *Discrete & Computational Geometry*, 42(4):517–526, 2008.

[100] Jiří Matoušek, Micha Sharir, and Emo Welzl. A subexponential bound for linear programming. *Algorithmica*, 16(4):498–516, 1996.

[101] Jiří Matoušek and Petr Škovroň. Removing degeneracy may require a large dimension increase. *Theory of Computing*, 3(1):159–177, 2007.

[102] Daniel Maurer and Christian Wieners. A parallel block LU decomposition method for distributed finite element matrices. *Parallel Computing*, 37(12):742–758, 2011.

[103] W. Gregory Medlin and James F. Kaiser. Bandpass digital differentiator design using quadratic programming. In *Proceedings of the IEEE International Conference on Acoustics, Speech and Signal Processing (ICASSP), New York, NY, USA*, volume 3, pages 1977–1980, 1991.

[104] Akira Mohri. A computational method for optimal control of a linear system by quadratic programming. *International Journal of Control*, 11(6):1021–1039, 1970.

[105] James A. Momoh, Rambabu Adapa, and Mohamed E. El-Hawary. A review of selected optimal power flow literature to 1993, part I. Nonlinear and quadratic programming approaches. *IEEE Transactions on Power Systems*, 14(1):96–104, 1999.

[106] Hiroyuki Mori and Senji Tsuzuki. A robust QP-based algorithm for power system state estimation. In C. E. Lin, editor, *Proceedings of the IASTED International Conference on High Technology in the Power Industry*, pages 130–134. ACTA Press, Calgary, Canada, 1991.

[107] John A. Muckstadt. A dual decomposition algorithm for solving mixed integer-continuous quadratic programming problems. Technical report, Air Force Systems Command, Wright-Patterson AFB, OH, USA, 1969.

[108] Thom Mulders. A generalized Sylvester identity and fraction-free random Gaussian elimination. *Journal of Symbolic Computation*, 31(4):447–460, 2001.

[109] Katta G. Murty and Santosh N. Kabadi. Some *NP*-complete problems in quadratic and nonlinear programming. *Mathematical Programming*, 39(2):117–129, 1987.

[110] George C. Nakos, Peter R. Turner, and Robert M. Williams. Fraction-free algorithms for linear and polynomial equations. *SIGSAM Bulletin (ACM Special Interest Group on Symbolic and Algebraic Manipulation)*, 31(3):11–19, 1997.

[111] Janardan Nanda, Dwarkadas P. Kothari, and Suresh C. Srivastava. New optimal power-dispatch algorithm using Fletcher's quadratic programming method. *IEEE Proceedings C (Generation, Transmission and Distribution)*, 136(3):153–161, 1989.

[112] D. L. Nash and G. W. Rogers. Risk management in herd sire portfolio selection: a comparison of rounded quadratic and separable convex programming. *Journal of Dairy Science*, 79(2):301–309, 1996.

[113] Søren S. Nielsen. Dense and sparse matrix classes using the C++ standard template library. *Computational Economics*, 14(1-2):47–68, 1999.

[114] Sven Nordebo, Ingvar Claesson, and Zhuguan Zang. Optimum window design by semi-infinite quadratic programming. *IEEE Signal Processing Letters*, 6(10):262–265, 1999.

[115] Pavel Okunev and Charles R. Johnson. Necessary and sufficient conditions for existence of the LU factorization of an arbitrary matrix, 2005. arXiv:math/0506382v1.

[116] K. A. Palaniswamy, Jaydev K. Sharma, and Krishna B. Misra. Minimization of load curtailment in power system using quadratic programming. *Journal of the Institution of Engineers (India). Electrical Engineering Division*, 65:213–218, 1985.

[117] Cornelis Van De Panne and Andrew Whinston. A comparison of two methods for quadratic programming. *Operations Research*, 14(3):422–441, 1966.

[118] Panos M. Pardalos and Georg Schnitger. Checking local optimality in constrained quadratic programming is *NP*-hard. *Operations Research Letters*, 7(1):33–35, 1988.

[119] David Poole. *Linear Algebra: A Modern Introduction*. Cengage Learning, Stamford, CT, USA, 2005.

[120] Alex Pothen, Horst D. Simon, and Kang-Pu Liou. Partitioning sparse matrices with eigenvectors of graphs. *SIAM Journal on Mathematical Analysis and Applications*, 11(3):430–452, 1990.

[121] Gerald F. Reid and Lawrence Hasdorff. Economic dispatch using quadratic programming. *IEEE Transactions on Power Apparatus and Systems*, PAS-92(6):2015–2023, 1973.

[122] John K. Reid. A sparsity exploiting variant of the Bartels-Golub decomposition for linear programming bases. *Mathematical Programming*, 24(1):55–69, 1982.

[123] Rolf Robleda. Eine dünnbesetzte Version des CGAL-Lösers für quadratische Programme. Master's thesis, ETH Zürich, Switzerland, 2008.

[124] R. Tyrell Rockafellar. Computational schemes for large-scale problems in extended linear-quadratic programming. *Mathematical Programming, Series B*, 48(3):447–474, 1990.

[125] R. Tyrell Rockafellar and Jonathan Z. Sun. A simplex-active-set algorithm for piecewise quadratic programming. In D.-Z. Du and Jie Sun, editors, *Advances in Optimization and Approximation*, pages 275–292. Kluwer Academic Publishers, Dordrecht, Netherlands, 1994.

[126] Donald J. Rose, Robert E. Tarjan, and George S. Lueker. Algorithmic aspects of vertex elimination on graphs. *SIAM Journal on Computing*, 5(2):266–283, 1976.

[127] Sartaj Sahni. Computationally related problems. *SIAM Journal on Computing*, 3(4):262–279, 1974.

[128] Sven Schönherr. *Quadratic Programming in Geometric Optimization: Theory, Implementation, and Applications.* PhD thesis, ETH Zürich, Switzerland, 2002.

[129] Jack Sherman and Winifred J. Morrison. Adjustment of an inverse matrix corresponding to a change in the elements of a given column or a given row of the original matrix. *Annals of Mathematical Statistics*, 20(4):620–624, 1949.

[130] Jack Sherman and Winifred J. Morrison. Adjustment of an inverse matrix corresponding to a change in one element of a given matrix. *Annals of Mathematical Statistics*, 21(1):124–127, 1950.

[131] Jae K. Shim. A survey of quadratic programming applications to business and economics. *International Journal of Systems Science*, 14(1):105–115, 1983.

[132] Petr Škovroň. *Abstract Models of Optimization Problems.* PhD thesis, Charles University, Prague, Czech Republic, 2007.

[133] Gilbert W. Stewart. *Introduction to Matrix Computation.* Academic Press, New York, NY, USA, 1973.

[134] Josef Stoer. Principles of sequential quadratic programming methods for solving nonlinear programs. In K. Schittkowski, editor, *Computational Mathematical Programming*, pages 165–207. Springer-Verlag, Berlin, Germany, 1985.

[135] Volker Strassen. Gaussian elimination is not optimal. *Numerische Mathematik*, 13(4):354–356, 1969.

[136] Bjarne Stroustrup. *Programming: Principles and Practice Using C++.* Addison Wesley, Boston, MA, USA, 2008.

[137] Leena Suhl and Uwe H. Suhl. A fast LU update for linear programming. *Annals of Operations Research*, 43(1):33–47, 1993.

[138] Wai S. Tang and Jun Wang. A discrete-time Lagrangian network for solving constrained quadratic programs. *International Journal of Neural Systems*, 10(4):261–265, 2000.

[139] William F. Tinney and John W. Walker. Direct solution of sparse network equations by optimally ordered triangular factorization. *Proceedings of the IEEE*, 55(11):1801–1809, 1967.

[140] Sivan Toledo. Locality of reference in LU decomposition with partial pivoting. *SIAM Journal on Matrix Analysis and Applications*, 18(4):1065–1081, 1997.

[141] Alan M. Turing. Rounding-off errors in matrix processes. *Quarterly Journal of Mechanics and Applied Mathematics*, 1(1):287–308, 1948.

[142] Peter R. Turner. Gauss elimination: workhorse of linear algebra. Technical report, Naval Air Warfare Center, Aircraft Division, 1995.

[143] Peter R. Turner. A simplified fraction-free integer gauss elimination algorithm. Technical report, Naval Air Warfare Center, Aircraft Division, 1995.

[144] Stephen A. Vavasis. Quadratic programming is in *NP*. *Information Processing Letters*, 36(2):73–77, 1990.

[145] Timothy Vismor. Matrix algorithms. Available on the internet https://vismor.com/download/Documents/Network_Analysis/ matrix_algorithms.pdf, 1990-2012.

[146] Hua Wei and Hiroshi Sasaki. Large-scale optimal power-flow based on interior-point quadratic programming of solving symmetrical indefinite system. In *Large Scale Systems: Theory and Applications*, pages 377–382, 1995.

[147] Emo Welzl. Smallest enclosing disks (balls and ellipsoids). *Lecture Notes in Computer Science*, 555:359–370, 1991.

[148] Frans Wessendorp. Degeneracy. Technical report, CGAL documentation.

[149] Frans Wessendorp. Update Uz. Technical report, CGAL documentation.

[150] Frans Wessendorp. Upper bounding. Technical report, CGAL documentation.

[151] Virginia V. Williams. Breaking the Coppersmith-Winograd barrier. Preprint, 2011.

[152] Philip Wolfe. A simplex method for quadratic programming. *Econometrica*, 27(3):382–398, 1959.

[153] Max A. Woodbury. Inverting modified matrices. Memorandum Report 42, Statistical Research Group, Princeton, NJ, USA, 1950.

[154] Mihalis Yannakakis. Computing the minimum fill-in is *NP*-complete. *SIAM Journal on Algebraic Discrete Methods*, 2(1):77–79, 1981.

[155] Yinyu Ye. Further developments on the interior algorithm for convex quadratic programming. Technical report, Department of Engineering-Economic Systems, Stanford University, CA, USA, 1987. Preprint.

[156] Ya-Xiang Yuan. A review of trust region algorithms for optimization. In *ICIAM '99 (Edinburgh)*, pages 271–282. Oxford University Press, Oxford, England, 2000.

[157] Guanglu Zhou, Kim-Chuan Tohemail, and Jie Sun. Efficient algorithms for the smallest enclosing ball problem. *Computational Optimization and Applications*, 30(2):147–160, 2005.

[158] Wenqin Zhou and David J. Jeffrey. Fraction-free matrix factors: new forms for LU and QR factors. *Frontiers of Computer Science in China*, 2(1):67–80, 2008.

[159] Wenqin Zhou, David J. Jeffrey, and Robert M. Corless. Fraction-free forms of LU matrix factoring. *Transgressive Computing*, pages 443–446, 2006.

[160] Karl-Heinz Zimmermann and Wolfgang Achtziger. On time optimal implementation of uniform recurrences onto array processors via quadratic programming. *Journal of VLSI Signal Processing Systems for Signal, Image, and Video Technology*, 19(1):19–38, 1998.

Nomenclature

The nomenclature is divided into three sections, which are *General*, *Violator Spaces*, and *Matrices & LU factorization*. Within a section the elements are ordered somewhat arbitrarily. As a rule of thumb, basic stuff is at the beginning, while more specialized notation is at the end. After a short description of the symbol we indicate the page number of its first use. We are confident that readers will find help by first choosing the appropriate section and then scanning for the symbol they are looking for.

General

Matrices & LU factorization

List of Algorithms